International Approaches to Securing Radioactive Sources Against Terrorism

edited by

W. Duncan Wood
Institute for Applied Science
Washington, DC
U.S.A.

and

Derek M. Robinson
Trilateral Group Limited
London, UK

 Springer

Published in cooperation with NATO Public Diplomacy Division

Proceedings of the NATO Advanced Research Workshop on
International Approaches to Securing Radioactive Sources Against Terrorism
Woodlands Park, Surrey, United Kingdom
4–6 November 2005

Library of Congress Control Number: 2008937991

ISBN 978-1-4020-9284-8 (PB)
ISBN 978-1-4020-9271-8 (HB)
ISBN 978-1-4020-9272-5 (e-book)

Published by Springer,
P.O. Box 17, 3300 AA Dordrecht, The Netherlands.

www.springer.com

Printed on acid-free paper

Internationa
Radioactive

6140081906

NATO Science for Peace and Security Series

This Series presents the results of scientific meetings supported under the NATO Programme: Science for Peace and Security (SPS).

The NATO SPS Programme supports meetings in the following Key Priority areas: (1) Defence Against Terrorism; (2) Countering other Threats to Security and (3) NATO, Partner and Mediterranean Dialogue Country Priorities. The types of meeting supported are generally "Advanced Study Institutes" and "Advanced Research Workshops". The NATO SPS Series collects together the results of these meetings. The meetings are co-organized by scientists from NATO countries and scientists from NATO's "Partner" or "Mediterranean Dialogue" countries. The observations and recommendations made at the meetings, as well as the contents of the volumes in the Series, reflect those of participants and contributors only; they should not necessarily be regarded as reflecting NATO views or policy.

Advanced Study Institutes (ASI) are high-level tutorial courses intended to convey the latest developments in a subject to an advanced-level audience

Advanced Research Workshops (ARW) are expert meetings where an intense but informal exchange of views at the frontiers of a subject aims at identifying directions for future action

Following a transformation of the programme in 2006 the Series has been re-named and re-organised. Recent volumes on topics not related to security, which result from meetings supported under the programme earlier, may be found in the NATO Science Series.

The Series is published by IOS Press, Amsterdam, and Springer, Dordrecht, in conjunction with the NATO Public Diplomacy Division.

Sub-Series

A.	Chemistry and Biology	Springer
B.	Physics and Biophysics	Springer
C.	Environmental Security	Springer
D.	Information and Communication Security	IOS Press
E.	Human and Societal Dynamics	IOS Press

http://www.nato.int/science
http://www.springer.com
http://www.iospress.nl

Series C: Environmental Security

CONTENTS

Contributors... vii

Abstract xi

Foreword xiii

SECTION I – Executive Summary...1

   1 Executive Summary of Discussions..................................... 3
       Dr. W. Duncan Wood

SECTION II – Security of Radioactive Sources 13

   1 A Historical Perspective and Recent Developments........................ 15
       Dr. Abel Gonzalez

SECTION III – High Consequence Radioactive Terrorism Scenarios 51

   1 High-Consequence Radioactive Terrorism Scenarios – Russia....... 53
       Dr. Leonid Bolshov

   2 High Consequence Radiological Terrorism Scenarios – UN........... 79
       Dr. Alex P. Schmid

   3 Radiological and Nuclear Terrorism................................. 99
       David Martin

SECTION IV – Coordinating Responses to Radioactive Terrorism........ 109

   1 Keeping the Terrorist Tragedies of Yesterday from
      Becoming the Terrorist Catastrophies of Tomorrow 111
       Dr. John Hnatio

SECTION V – Future Security of Radioactive Sources 131

   1 Lessons Learned, Weak Points and Future Improvements............ 133
       Ivan Gorinov

   2 Protecting Major Public Events Against Nuclear
      Radiological Terrorism 145
       Dr. Klaus Duftschmid

   3 Securing Radioactive Sources Against Terrorism in Georgia 161
       Levan Gogua

SECTION VI – Uncorking the Bottlenecks 167

1 Breaking Down Barriers to Cooperation Between
   Governments, Civil Society and Business 169
   Michael McKay

APPENDIX I 183

1 Searching for Hidden Radioactive Sources – Experience
   from Exercises in Poland 185
   Genowefa Smagala

2 Letter from US State Department in Support of
   the Workshop Initiative 191

3 Letter from Baroness Thatcher in Support of
   the Workshop Initiative 193

APPENDIX II – Background Papers from International
   Approaches to Nuclear and Radiological Security
   Conference 2002 (IANRS) 195

1 Chairmen's Summary and Evaluation 197
   John W. Wood and Dr. Evgeny Velikhov

2 Findings of the Working Group on Trends in Illicit Trafficking 205

3 Findings of the Working Group on Radiological Threat
   Reduction 211

4 The Russian Academy of Sciences and International
   Coperation 217
   Nikolai Platé

5 VECTOR and International Cooperation 221
   Lev S. Sandakhchiev

6 Address to the International Approaches to Nuclear and
   Radiological Security Conference, 2002 225
   Baroness Thatcher of Kesteven

7 International Nuclear Material Protection and Cooperation
   Sites 229

CONTRIBUTORS

Workshop Co-Directors

Dr. Evgeny Velikhov, President, The Russian Research Centre, Kurchatov Institute, Russian Federation. Director, Institute for Applied Science Inc., Moscow, Russia

John W. Wood, Chairman, Trilateral Group Ltd. and Institute for Applied Science Inc., UK

Participants

Ali Aygun, Head of Department, Customs Undersecretariat, General Directorate of Customs Enforcement, Turkey

Jackie Boesinger, Sales Leader, GE Energy – Reuter-Stokes Inc., USA

Leonid Bolshov, Director, Nuclear Safety Institute (IBRAE), Russian Academy of Sciences, Russian Federation

Robin Booth, Security Consultant, Trilateral Group Ltd., UK

Todd Brethauer, Senior Scientist, Combating Terrorism Technology Support Office – TSWG, Department of Defense, USA

Jack Caravelli, Senior Visiting Fellow, UK Defence Academy, UK and Former Senior Adviser, Office of Strategic Planning and Programs, National Nuclear Security Administration (NNSA), Department of Energy, USA

Finn Chemnitz, Defense Policy and Planning Division, NATO Headquarters, Belgium

Dimitry Danko, Vice-Chief, Fissionable and Radioactive Materials, Federal Customs Service, Russian Federation

Tim Doran, Program Manager, HM Revenue & Customs-Cyclamen, UK

Klaus Duftschmid, Professor for Radiation Protection, Technical University Graz. Consultant to Department of Safeguards and Technical Cooperation, International Atomic Energy Agency (IAEA), Austria

Al Frymire, President, TSA Systems Ltd., USA

Abel Gonzalez, Senior Advisor, Nuclear Regulatory Authority, Argentina. Former Director – Division of Radiation and Waste Safety, International Atomic Energy Agency (IAEA), Argentina

Ivan Gorinov, Head, Division of Nuclear Material and Physical Protection, Nuclear Regulatory Agency, Bulgaria

Levan Gogua, Deputy Director, Service for Nuclear and Radioactive Safety, Georgia

Matthew J.F. Healy, Lecturer, Cranfield University, Defence Academy of the United Kingdom, UK

John Hnatio, Executive Director, Institute for Complexity Management. Former Professor of Strategic Leadership and Decision-making, National Defense University, USA

Ian Kerr, Managing Director, William Bain Fencing Ltd., UK

Nikolay Kravchenko, Vice-Chief, Directorate of Special Techniques and Automation of Customs Technologies, Russian Federation

Glen Levis, Senior Analyst, US Government Accountability Office – NRE, USA

Susan Lubomirski, Trilateral Conference Organiser. Director, Spice Ltd., UK

David Martin, Director, Strategic Defense Programs, Tetra Tech EC Inc., Former Program Manager – Second Line of Defense, International Material Protection and Cooperation Program, Department of Energy, USA

Jim McColm, Cyclamen Intelligence Manager, HM Revenue & Customs-Cyclamen, UK

Michael McKay, CEO, McKay's, International Corporate Communications & Public Affairs Management, Switzerland

Philip Petersen, Director, Institute for Applied Science, USA

Derek Robinson, Director, Trilateral Group Ltd., UK

Alex P. Schmid, Senior Crime Prevention and Criminal Justice Officer, United Nations Terrorism Prevention Branch, United Nations Office for Drug Control and Crime (UNODC), Austria

Genowefa Smagala, Project Leader, Central Laboratory for Radiological Protection, CLOR, Poland

David Soumbadze, Deputy Chief of Mission, Embassy of Georgia to the United States of America, Canada and Mexico, USA

David Thornbury, Cyclamen Specialist, Threat Reduction – Atomic Weapons Establishment (AWE), UK

Anna Tsiporkina, Institute for Applied Science Liaison to Russian Academy of Sciences. President, Eurasian Technologies, Kazan, Russian Federation

W. Duncan Wood, Research Director, Institute for Applied Science, USA

Rob York, Distinguished Research Staff, Oak Ridge National Laboratory, Department of Energy, USA

Workshop hosted by Trilateral Group Ltd, London
Supported by General Electric (GE), Tetra Tech Inc.
and TSA Systems

ABSTRACT

International Approaches to Securing Radioactive Sources Against Terrorism

As a result of arms control efforts over the past 50 years, nuclear material is subject to strict national controls and tough international treaties. But there are still almost no controls, other than a voluntary International Atomic Energy Agency code of conduct, on the sorts of radiological sources used to make radiological dirty bombs. Radiological sources are used all over the world for a wide range of peaceful purposes, including smoke detectors, medical devices, meteorology, mining and thermoelectric generators. There are at least 8 million identified radiological sources worldwide. Their small size, portability and high value make them vulnerable to misuse and theft: the IAEA reported 272 cases of illicit trafficking in sealed radioactive sources between 1993 and 2002.

The IAEA estimates that 110 countries worldwide still fail to impose adequate controls. The time is ripe for an international convention and treaty on the safety and security of radiological sources.

This two-day workshop, held at Woodlands Park, Surrey, UK in November 2005, was designed to enhance cooperation and assistance between NATO and Partner countries in support of International Atomic Energy Agency (IAEA) efforts to secure radioactive sources against the threat of terrorism and also to support the Security Agenda at the International Radiation Protection Association Congress in Buenos Aires in 2008.

The workshop assessed likely scenarios in which radioactive sources could be used as terrorist weapons and identified means to enhance operational responses to high consequence radiological terrorism such as a radiological dispersion device or 'dirty bomb'. In addition, the workshop assessed indicators and warnings of clandestine radiological acquisition and development and also focused on technical cooperation in the development of new technologies for detection, monitoring, interdiction and enhanced security of radioactive sources. Finally, the workshop considered shortfalls in existing legislative and regulatory controls over sealed sources and examined ways to improve the implementation of key aspects of regulatory regimes such as: licensing and inspection; tracking sources; registries of source users; international and national source inventories; identification and recovery of orphan sources; transport security.

Supported by the NATO Security Through Science program, the workshop is part of the ongoing program of activities established in 2002 at the International Approaches to Nuclear and Radiological Security Conference (IANRS) in London with support from the US Department of Energy's National Nuclear Security Administration (NNSA) and the Russian Federation's Ministry of Atomic Energy and the Kurchatov Institute.

Other NATO Workshops in this IANRS series include:

- Emerging Threats to Energy Security and Stability, January 2004

- International Approaches to Illicit Trafficking: Detection, Deterrence and Interdiction, September 2004

- Black Sea Security: Enhancing Regional Cooperation to Counter Illicit Trafficking, June 2006

- International Approaches to Counter Next-Generation Improvised Explosive Devices, April 2008

FOREWORD

John W. Wood
Workshop Co-Chairman. Chairman, Trilateral Group Ltd., UK

Chairman Velikhov, distinguished speakers and participants, friends and colleagues. It is my great pleasure to welcome you to this NATO workshop on Radiological Security here at Woodlands Park.

The concern of this workshop, Radiological Security, moves ever higher up the list of global security priorities. This fact, I know, puts you under greater pressure and makes greater demands on your schedules than ever before. So we are particularly grateful that you have been able to make the time to be here this weekend. Many of you have travelled great distances to join us.

This workshop springs from the work of the inaugural conference on International Approaches to Nuclear and Radiological Security sponsored by the US and the Russian governments and the Research Centre the Kurchatov Institute in Moscow, which we held in London in October 2002.

That conference produced a powerful set of actionable recommendations, one of the most important of which was that each working group of the conference be kept in place and its work continued, and reviewed on an annual basis.

In this regard I would like to pay a special tribute to Charlotte Wood who devoted the last year of her life to bringing that seminal conference to fruition. And whose extraordinary organisational skills, tireless attention to detail, exceptional gift for diplomacy and genuine and unfailing cheerfulness contributed more than anything else to make it an outstanding success of International Cooperation.

Towards that end we began collaboration with NATO to sponsor annual workshops aimed at progressing this work. Last year we held the

NATO workshop on Illicit Trafficking at Norton Manor, about 50 miles from here. This weekend's conference follows logically from that.

The Radiological working group of the 2002 conference, so ably chaired by Professor Leonid Bolshov, produced important recommendations.

This weekend's workshop was motivated by Abel Gonzalez's call at last year's meeting to organise a follow up one to enhance international collaboration to support the IAEA's efforts to secure Radioactive Sources. The deliberations of that workshop produced further useful recommendations, principal among these were:

1. The establishment of a NATO Advisory Group on Illicit Trafficking to provide NATO with an enhanced counter-trafficking support capability.

2. A follow up NATO workshop on The Security of Radiological Sources in support of the IAEA Division of Radiation and Waste Safety meetings in Bordeaux 2005.

3. The establishment of an International Nuclear and Radiological Security Training Institute – using NATO member and partner capabilities including The Kurchatov Institute's firsthand experience of Nuclear and Radiological crises.

4. Adapting the US Counter Narcotics Joint Agency Task Force (JIATF) concept of wider counter-trafficking/counter terrorism objectives.

5. The establishment of a Black Sea/Caucasus Regional Counter Trafficking Centre.

6. Include Industry in counter trafficking, planning and training.

7. The media have become the medium – information preparedness and crisis communication is now a pre-condition of effective security response.

8. Follow on activity on Information Preparedness.

9. To use NATO workshop publications as a follow up tool.

10. For the NATO organisers to maintain close communication with members of the working group in order to facilitate the sharing of information among participants, and follow up with regard to action items.

Since last year the work has become even more important. It is a paradox that while Radiological Security has generally received less attention and urgency than Nuclear Security, yet the threat-threshold probability of the use of a Radiological device is considerably higher than that of a nuclear device. Why is this? The reason is simple: nuclear materials are fewer, better State controlled, and more difficult to use. A nuclear device would, it is true, cause more physical damage, but in the age of terrorism it is not physical damage that is the point.

Terrorism is a tactic. Its target is not physical entities, but the minds of those possessing those entities. Terrorism is war on the mind. Given that in the popular mind the distinction between Nuclear and Radiological devices are largely unappreciated, it follows that radiological devices because of their ease of acquisition and use, and their equivalence as a weapon of terror, will be the weapon of choice. Thus, because the probability of the use of radiological devices is considerably *higher* than the probability of use of a nuclear device it is clear that radiological security needs to move up the hierarchy of priorities. A radiological device will not produce as big a bang against property and populations as a nuclear device, but it can produce an equally big bang against the mind, and it's easier to do.

This raises many issues some of which we will deal with here, some of which will be the subject of future workshops: these include the role, responsibilities of, and levels of knowledge of the media, and the need to achieve a differentiation between nuclear and radiological devices in the media's reporting, and in the public mind. And doing it well before an event.

Therefore, what we are seeking in this workshop and others, is the development of a stronger and more cohesive thinking capability on these issues: and more effective cross-cutting working relationships between relevant entities and actors within the new paradigms of cooperation and communications demanded by the security environment post 9/11.

Speaking at last year's workshop, Chris Donnelly from NATO posed the dilemma well:

"In all our countries, as in the corridors of NATO and EU, the current pace of events faces civilian and military staffs with enormous burdens of overwork."

"There is no longer enough time to deal with everyday problems and find enough time for conceptual thinking. After all, the scope of change being forced upon institutions is the greatest it's ever been in peacetime."

"Consequently there is a great need to generate ideas, stimulate thinking and debate on the issue of preventing radiological terrorism, as on the issue of all aspects of security reform. We need to break down boundaries between different elements of the security establishment and to expand the frontiers of what is considered 'security'. There is an equal need to increase the strength of the 'security community' – the body of military and especially of civilian personnel competent in the new security issues and capable [a] of filling posts in national and international institutions; and [b] educating the population to understand the new needs of security so as to ensure their support through the democratic process."

"This is both a short-term and a long-term requirement. The think tank is now the interface between the brainpower of the academic community on the one hand and the overworked policy community, which needs intellectual support, on the other. It is also the cradle for educating the new generation to deal with the security threats of today and tomorrow".

These workshops are a partial response. The hope is to create what are in effect virtual think tanks, which meet once a year, but which work and think collaboratively in between.

In some senses the structure of this workshop evidences the new paradigm. For its principal motivators are, first, the original conference sponsored by two governments, second the steadfast support of NATO, a military alliance which increasingly seeks to make an intellectual contribution to the emerging paradigm created by uncontrollable proliferation of technology, the security gap between rich and poor countries, and the information revolution. Third the IAEA, it is a specialised operational arm of a truly global organisation, the United Nations. It is instructive that in its report on Nuclear Security tabled at the general conference of the IEAE it specifically spelt out that "the agency has continued to seek liaisons, collaboration and coordination with other regional, transnational and international organisations in carrying out its mandate".

The fourth and final strand of this workshop is the private sector: we benefit from the generous support of three major corporations, all with specialised knowledge in relevant fields: Tetra Tech, GE Energy and TSA Systems.

So as we begin our work I would particularly like to congratulate Dr. Jack Caravelli who had the foresight to create and sponsor the original conference in 2002 when he so successfully ran the international MPCA program at the US Department of Energy and who has taken time to be with us today. I would like to thank my Co-Chairman Dr. Evgeny Velikhov, Director of the Kurchatov Institute, Science Advisor to President Putin, long term member of the Russian Security Council, fellow founding Director of the Institute of Applied Science, one of the world's most distinguished Nuclear Physicists, and a tireless traveller and worker in the interests of Nuclear and Radiological progress and global security.

Evgeny has one the most hectic travel schedules of anybody I know so it is a particularly great pleasure that he has been able to be here to Chair our proceedings.

I'd like to thank Chris Donnelly from NATO who, with typical prescience, has led the creation in NATO of an intellectual and analytical capability to embrace the new security issues in their widest and most profound sense. It was Chris who forged our working relationship with the NATO Advanced Science Program and he is, in a sense, the father of this entire enterprise. I'd like to thank Abel Gonzalez of the IAEA for his wisdom and steady support and, not least, his willingness to travel all the way from Buenos Aires to be with us and further the ends of this workshop. Finally, I would like to thank David Martin of Tetra Tech who is not only lending his expertise to panel number two, but also acts as one of the three major corporate supporters. I'd like to thank the experts from GE Energy for attending this conference and acting as corporate supporters, and also the experts from TSA Systems for also participating and acting as corporate supporters. I'd like to thank all three for being so generous with their intellectual and financial support.

And not to leave parliamentarians out, we owe Congressman Curt Weldon a vote of thanks for he has steadily supported our work. Curt had intended to be here but a tricky scheduling of a Congressional delegation through RAF Northolt, has in the end meant that he had to miss the weekend. And finally I'd like to thank Lady Thatcher who over many years has lent us her considerable moral and intellectual support to our work.

As Lady Thatcher said in her speech at the inaugural conference in 2002 "Today we are making up for lost time, but have no doubt the time must be made up. Because though we might wish for the best we must prepare for the worst, and that is why your conference could not come at a more crucial time. My friends, what was true two years ago is even more true today. We are now forging a partnership which will bring dividends both for us and for future generations. You serve not only your own nations but all nations by your efforts". Lady Thatcher also wrote last year "I must congratulate you for the central role which you and your colleagues continue to play in encouraging and keeping this vital community together."

So welcome again, my friends. Let us now set to work in honouring that trust.

SECTION I

EXECUTIVE SUMMARY

1   EXECUTIVE SUMMARY OF DISCUSSIONS

Dr. W. Duncan Wood
Research Director, Institute for Applied Science, USA

Overview

The NATO Advanced Research Workshop on International Approaches to Securing Radioactive Sources Against Terrorism took place in November 2005 at Woodlands Park, UK and was organized by the Trilateral Group with additional assistance from the Institute for Applied Science, Tetra Tech EC, TSA Systems and GE. The Co-Directors were John Wood, Chairman of Trilateral Group and Dr. Evgeny Velikhov, President, the Russian Research Center, Kurchatov Institute, Moscow.

The workshop provided a comprehensive historical overview of regulatory control over radiological sources and then identified the key regulatory achievements and shortfalls that have characterized international efforts to control radioactive sources since the terrorist attacks of September 11, 2001. The workshop participants assessed high consequence radiological terrorism scenarios and identified means to enhance operational responses to prevent and respond to high consequence radiological terrorism. There was a session focused on strengthening technical cooperation in the development of new technologies for detection, monitoring, interdiction and enhanced security of radioactive sources. The final panel sought to identify specific obstacles to be overcome and actions needed at national and international level to ensure cradle to grave control of radioactive sources.

At the public diplomacy level, the workshop served as a forum to enhance public–private sector cooperation and assistance between NATO and Partner countries in support of International Atomic Energy Agency (IAEA) efforts to secure radioactive sources against the threat of terrorism.

W.D. Wood and D.M. Robinson (eds.), *International Approaches to Securing Radioactive Sources Against Terrorism*,
© Springer Science+Business Media B.V. 2009

The workshop also helped to familiarize national radiological experts with the work of the NATO WMD Center.

Supported by the NATO Security Through Science program, the workshop was part of the ongoing program of activities of the Radiological Security Working Group established in 2002 at the International Approaches to Nuclear and Radiological Security Conference in London with support from the US Department of Energy's National Nuclear Security Administration (NNSA) and the Russian Federation's Ministry of Atomic Energy (MinAtom) and the Kurchatov Institute.

Conclusion: The Time Is Ripe for a Binding International Treaty to Secure Radioactive Sources

The relative ease of access to radioactive sources compared to nuclear materials makes the probability of their use by terrorists extremely high.

The terrorist events of September 11, 2001 marked a change in the nature of conflict and highlighted the international community's need to develop corresponding approaches to radiological security. There are considerable shortfalls in the control of radioactive sources and the time is right to press for a binding international treaty to implement a harmonized, effective and sustainable international regime for the safety and security of radioactive sources.

Findings: Radiological Sources Are Vital, Vulnerable, Misunderstood and Largely Unregulated

Vital

- Radiological sources are ubiquitous and invaluable in modern society.
- Sealed sources are used all over the world for a wide range of peaceful purposes including: smoke detectors, medical devices, radiotherapy, meteorology, mining and thermoelectric generators.
- There are 192 states in the world – all with radiation safety and security problems.
- At least 8 million identified sources worldwide and 2 million licensed sealed sources in the USA alone.

Vulnerable

- Radiological sources are easy-to-conceal, highly portable and poorly controlled.
- A radiological dirty bomb is low-tech, low-cost, and easy-to-deploy. Suicide terrorism makes assembly and delivery easy.
- The relative ease of access to radiological sources as compared to nuclear material makes the probability of their use by terrorists higher.
- It is relatively easy for terrorists to deploy sealed radioactive sources as terrorist weapons. A dispersion device combining highly radioactive material used in radiography such as cobalt-60, cesium-137, or iridium-192 packaged with conventional explosives would be highly effective in causing a high number of casualties and extensive economic damage. Even deploying a single canister of radioactive material without any explosive would be an effective terrorist weapon.
- The small size, portability and relatively high value of sealed sources makes them vulnerable to misuse and theft – the IAEA reported 272 cases of illicit trafficking in sealed radioactive sources between 1993 and 2002.

Misunderstood

- Radiological terrorism poses significantly different security threats from nuclear terrorism but has a similar psychological terror effect.
- The primary impact of a dirty bomb is economic and psychological – not mass casualties and physical destruction.
- Terrorism is war on the mind. The public does not differentiate between nuclear and radiological terrorism. So education on the difference is vital.
- To counter post 9/11 terrorism we need to break down boundaries between different elements of security establishment and expand the frontiers of what is considered 'security'.

Largely Unregulated in Comparison with Nuclear Material

- Nuclear material is subject to strict national controls and tough international treaties. In contrast, radiological sources are only subject to a voluntary international code of conduct.

- There are major shortfalls in both international and national legislative and regulatory controls of sealed radioactive sources. Despite IAEA efforts to improve national regulatory infrastructures, the IAEA estimates that 110 countries worldwide still fail to impose adequate controls over sealed sources. Moreover, there are significant control shortfalls even in the countries with relatively strong regimes. The European Union's member states lose up to 70 sealed radioactive sources annually, while the US Nuclear Regulatory Commission estimates that up to 250 sealed sources or devices are lost or stolen in America every year. As a result, the worldwide number of sealed sources in use, lost or stolen is unknown but the number of 'orphan' sources worldwide is generally estimated to be in the thousands.

Discussions

1. International Efforts to Regulate Radiological Sources Only Began in the 1990s and Are Still Not Backed Up by a Treaty

Unlike the international regime to control nuclear material which began immediately after World War II, international efforts to control radiological sources only began in earnest after the 1988 radiological accident in Goiania Brazil. The Goiania accident highlighted all the dangers inherent in poor control: an unsecured radioactive source from a medical facility containing 93 g of caesium 137 was stolen from a junk yard by scrap scavengers who then bust it open. As a result of this minor theft, 249 people and 85 houses were contaminated; 112,000 people needed to undergo radiation monitoring; 14 people were treated for radiation exposure and four people died within four weeks. More than 5,000 m^3 of radioactive waste had to be treated at a cost of approximately $30,000/m^3. The overall damage to the economy from this single accident is estimated at $36 million.

Ten years later, in 1998, the first IAEA Conference on the Safety of Radiation Sources and Security of Radioactive Materials was held in Dijon, France. Ironically, on September 10, 2001, the IAEA Board approved an International Action Plan on Safety and Security of Radioactive Sources. The next day on September 11, 2001 Al-Qaeda launched four simultaneous passenger plane suicide terrorist attacks on New York and Washington, DC.

The IAEA identified three plausible and distinctive threat scenarios, requiring three distinct solutions:

1. Detonating improvised nuclear devices
2. Sabotaging nuclear facilities
3. Dirty Bombs using radioactive materials

In 2002 the London Conference on International Approaches to Nuclear and Radiological Security (IANRS) identified distinctive requirements for the security of radiological sources and in 2003 and 2005 the IAEA hosted follow-on Conferences on Security of Radioactive Sources.

These conferences resulted in general agreement on the need to:

1. Locate, Recover and Secure Radioactive Sources Still at Large (orphan sources)
2. Ensure Global and Sustainable Control of Radioactive Sources

Technical consensus and guidance on these issues takes the form of four documents issued by the IAEA between 2003 and 2005:

1. "Categorization of Radioactive Sources"
2. "International Catalogue of Sealed Radioactive Sources and Devices"
3. "Code of Conduct on the Safety and Security of Radioactive Sources"
4. "Guidance on Import and Export of Radioactive Sources"

Considerable progress has been made in securing radioactive sources through the implementation of regional multilateral and bilateral initiatives such as the Tripartite Initiative between Russia (MinAtom), USA (the Department of Energy) and the IAEA to secure radiological sources in the states of the Former Soviet Union.

Despite this progress, whereas nuclear materials are subject to strict national and international controls, the only international regime for radioactive sources is the IAEA's voluntary code of conduct. Consequently, the IAEA estimates that 110 countries worldwide still fail to impose adequate controls over sealed sources and there are significant control shortfalls even in the countries with relatively strong regimes.

2. Radioactive Terrorism Scenarios and Responses

The IAEA Illicit Trafficking Database records a total of 665 cases of nuclear and radiological trafficking between 1995 and 2004 and case data from Eastern Europe suggests that the total number of cases is considerably higher due to underreporting. There have been several documented attempts of terrorists attempting to use radioactive substances. For example, in November 1995, Chechens threatened to detonate containers of radioactive caesium in a Moscow park, and in May 2002, the US Government arrested Jose Padilla (aka Al-Mudjahir) and charged him with planning to detonate an Improvised Radiological Device in Washington.

Four types of dispersion were identified:

1. A radiation dispersion device (RDD) in the form of a conventional explosive attached to radioactive material
2. A silent dispersal of radioactive materials in the air, the water or on the soil, e.g. by aerosol, dilution or dusting
3. A stationary radiological emission at a place where people reside (sometimes called a Radiological Emission Device)
4. An attack on a site containing radioactive materials and its dispersal there

Five types of consequences of a radiological attack were also identified:

1. Health impact on human beings exposed to radioactivity
2. Socio-psychological impact of the radioactive attack on direct and indirect victims, including changed behaviour of impacted segments of society
3. Political impact (including loss of confidence in government)
4. Economic impact (caused by denial of access in contaminated zone on the one hand and costs of cleaning up the polluted areas on the other)

5. Environmental impact

The public perception of the impact of radioactive disasters is far worse than the actual damage. Consequently Radiological Dispersal Devices (RDDs) are disproportionately effective weapons of terror. Figures from the Russian Nuclear Safety Institute (IBRAE) indicate that while the actual number of initial deaths from the reactor failure at Chernobyl is 31, the general public estimate in Russia of the death toll is 40,000. Similarly the number of subsequent deaths from Chernobyl is calculated at 60–80, yet the public estimate is 253,000.

3. Coordinating Responses to Radioactive Terrorism

The key task is to ensure that terrorist tragedies do not turn into catastrophes. It is clear that errors in the response system can lead to severe consequences for the population. The indirect consequences of inadequate management of radiation risks can have much greater impact than the direct losses. Therefore, we need to develop next-generation disaster preparedness decision support systems to aid decision-making. These systems need to be supported by effective computer models to predict the radiological consequences of a terrorist event. For example, a 3-Dimensional Transport Model of radiological pollution in urban conditions has been developed by the Russian Nuclear Safety Institute which provides timely analysis of predicted air contamination levels throughout the city.

4. Technical Cooperation

The discussions highlighted the effectiveness of border/portal monitoring using radiation detectors as an effective means to combat illicit trafficking in radiological sources. According to Russian Customs Service, 95% of sources are discovered using fixed and hand-held devices and only 5% through document checks and intelligence gathering.

The discussions also highlighted the need for technical cooperation on other aspects of radiological source security including:

- Geo-spatial registration of data and sources
- Data communication and display for incident command
- Low-false alarm rate
- Quick and reliable isotope identification

- Decontamination methods
- Public confidence and reassurance
- Training

There is considerable international technical cooperation on issues such as training, and installation of portal monitoring upgrades, but there are still too many failures of border controls. One way forward that was proposed is to develop an international Design Basis Threat standard for radiological sources similar to the security standard established for nuclear material in order to improve to provide "cradle to grave" control over such sources.

Action Items

The workshop produced several action items, principal among these are the following:

- The key issues for effective control over radiological sources remain simple to formulate and resolve.
- Prevalence of radioactive sources around the world. Solution – internationalize control.
- "Orphan sources", many radioactive sources are unaccounted for. Solution – help to find them and regain control.
- Relinquished Control, most radioactive sources are not well secured. Solution – impose international prescriptive regulatory requirements for ensuring control.
- Isolated Manufacturers. Solution – support the recently created association of manufacturers.
- Unconventional Sources. Solution – exert international pressure on countries that produced and abandoned these sources.
- Information preparedness/public and media education is essential to prevent and mitigate radiological terrorism.

The time is ripe for:

- International binding obligations to recover orphan sources.
- International binding obligations to ensure that the provisions of the IAEA Code of Conduct are followed by all.

- An International Convention on the Safety and Security of Radiological Sources to ensure that states make binding commitments to implement comprehensive cradle to grave security.
- States need to develop their own internal corresponding new control regimes and response mechanisms to counter radiological terrorism.

SECTION II

SECURITY OF RADIOACTIVE SOURCES

1   A HISTORICAL PERSPECTIVE AND RECENT DEVELOPMENTS

Dr. Abel Gonzalez
Senior Advisor, Nuclear Regulatory Authority, Argentina, Former Director, Division of Radiation and Waste Safety, International Atomic Energy Agency (IAEA), Argentina

Pre-History

My intention is to give a panoramic view of where we came from to where we are today. Let me start with a prehistory to give a context to the history. The prehistory lasted, surprisingly enough, until the 1990s. All international standards took security of radioactive sources for granted. The basic concept was that lax security situations should be prevented. There was a brief moment of enlightenment in the 1970s where there were international standards issued to cover the security of thermo-generators operated by high radioactivity sources. But I challenge you to find a copy of this in any of your libraries. Even at the International Atomic Energy Agency (IAEA) in Vienna, there is only one remaining. It is interesting because the Thermo generators follow many of these standards. But when the Soviet Union was disbanded, then the control was lost. One thing is certain – making history carries a lot of risk.

W.D. Wood and D.M. Robinson (eds.), *International Approaches to Securing Radioactive Sources Against Terrorism*,
© Springer Science+Business Media B.V. 2009

1988: Awakening

I would say that the real history of Radioactive Security started in 1988. What brought the problem to light for everyone was the accident in Goiânia, Brazil on 13 September 1987. All of us know that this was the first case where the loss of security produced a problem. Let me recall the details of Goiânia:

- There was an unsecured caesium 137 source in a radiological clinic.
- Scrap scavengers broke in, stole it and moved it to junkyard.
- The source capsule was ruptured and disposable insoluble caesium chloride was released.
- The city of Goiânia was contaminated.
- Fourteen people were overexposed – four died within four weeks.
- 112,000 people were monitored, of which 249 were contaminated.
- Eighty five houses were contaminated and hundreds of people were evacuated.
- Over 5,000 m^3 of radioactive waste was produced.

Now, apart from the first two items listed above, all the rest apply to what happens in a terrorist situation. The first two items were just a case of someone doing the wrong thing. And I should like to remind you that 93 g of a powder, similar in consistency to talcum powder, in a 2 in. diameter cylinder produced all the subsequent problems.

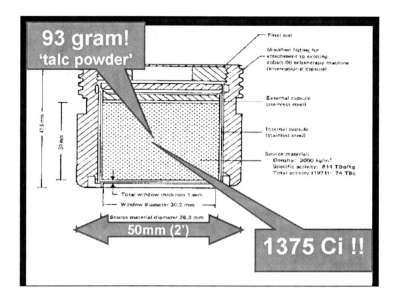

In the 1990s there was an international reaction to this problem. The international standards were approved by the board of the International Atomic Energy Agency in 1996, and it was the first time that the security of radioactive sources was introduced explicitly as a regulatory obligation into international standards.

1998 became what I call the era of reason leading to an understanding of the issues. The first conference, the conference of Dijon, was the first time that the issues were discussed by the scientific community. There were two main messages from Dijon: the first was that keeping radioactive sources under control is a serious international challenge; secondly, and most important, countries should undertake international obligations that guarantee proper control of radioactive sources. This was the main message from Dijon, and up to now, it has not been realized.

In 1999 we publicised the issues in an article in an IAEA bulletin, which is still available on the web and the issues now are still the same. This was the first time that it was published to decision-makers. (http://www.iaea.org/Publications/Magazines/Bulletin/Bull413/article1.pdf).

The year 2000 saw action in the regulatory area. There was a big meeting in Buenos Aires with all the regulators involved with this issue. And out of Buenos Aires came two main messages:

- First to the national regulators: you have a serious unresolved problem of security of radioactive sources, which incidentally is your responsibility. What do you plan to do about it?

- Second to the international community: the time is ripe for a serious global commitment and for an international action plan.

The international action plan was prepared and, ironically, on September 10, 2001 the action plan was approved. One day later the attack on the World Trade Center happened. There was a 24 hour difference. It was unbelievably ironic.

Of course 2001 was the year of terrorism and confusion. There were five ingredients for chaos. First, public concern and media zeal – the media want to know, the media want action. Second political demands: Politicians who were ignoring the problem until that moment, wanted a 24 hour solution. Third, lack of strategic planning: we had to recognize that there was no strategic plan at that moment. Fourth, the big budget effect: sometimes too much money is a bad thing, because people want to spend that money. Finally, there were the stumbling blocks of personal ambitions.

Initially there was clarity. Three diverse plausible scenarios were presented in a crystal clear post-September 11 document of the IAEA Board of Governors. It recognized that detonating improvised nuclear devices, sabotaging nuclear facilities, and the misuse of radioactive materials (dirty bombs) were three distinct, different problems.

But quickly the clarity degenerated into confusion. In the beginning of 2002 there was an obscure document from the same IAEA Board of Governors, which scrambled the approaches to solving the three distinct problems. This was not helpful. I think it was a big mistake of the agency. The security of nuclear weapons and materials, of nuclear installations, and of radioactive materials are three distinct issues. But they were scrambled! And nuclear security experts were playing the role of radiological security experts, and vice versa. The result was big problems for both sides.

In 2003 came the Renaissance. In March 2003 the IAEA held the Hofburg conference, which was one of the biggest events to address this topic. Nearly a thousand people from all over the world discussed the problem, and there were clear directions in the area of productivity that came out of Hofburg. The first, to locate, recover and secure powerful radioactive sources still at large. The second, to ensure global and sustainable control of radioactive sources.

This second point recognized, for the first time, that the issue was not only to solve the mess that we had at that moment but to avoid the same problem in the future. And in June 2003 the time was ripe for political understanding. At the G8 summit at Evian on France, these world's most powerful people recognized for the first time that there was a problem.

The Evian Declaration

You remember that the G8 Summit ended with a very important declaration. It was very important because Mr. Putin signed up to the declaration. Let me remind you that before that G8 meeting, the official position of Russia was that this important problem was not an IAEA problem – it was a Soviet Union problem. And this was the position going back as far as the Brezhnev era.

After Evian the position changed dramatically and there was enormous support from the Russian government. In July 2003, we reached a key international technical consensus: what was the meaning of dangerous radioactive sources? And the document IAEA Tech Doc 134 "Categorisation of radioactive sources" defined the meaning of 'dangerous radioactive sources' and gave us examples.

Code of Conduct

September 2003 saw a focus on what was important – for the first time it was realised that this problem would not be solved unless there was a strengthening of National Infrastructures for control. This was a very important issue for many countries around the world, and the conference at Rabat in Morocco made this very clearly understood.

But the biggest outcome was the adoption of the Code of Conduct. What is the Code of Conduct? I used to liken it to the Catechism for Christians: if you don't behave well you will go to the inferno but nothing will happen to you on earth. The Code of Conduct is not a legally binding commitment.

However, the political commitment is very big. Seventy-two states have signed up to the Code of Conduct. Whether they will comply or not is a different matter. But at least the signatures come from the top.

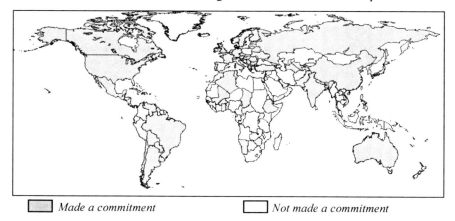

▨ *Made a commitment*   ☐ *Not made a commitment*

The Code provides a strategy that we have called security from cradle to grave – throughout the life-cycle of the radioactive material. More important, just a few months ago, there was an international agreement on import and export controls. And in three weeks time the first review of this will take place in Vienna. It was a very important agreement, and it was very difficult to achieve.

At the same time a NATO Illicit Trafficking Workshop took place with a recommendation to enhance international co-operation.

Bordeaux, 2005

The road from Dijon to Bordeaux was long and complicated. But this shows that there was political commitment driving it, which is very important.

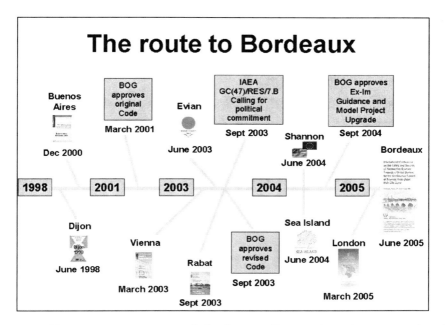

I will try to cover some of the issues discussed at the International Conference at Bordeaux. For the first time, the inventories of radioactive sources from various countries were presented. The French, for example, presented a very comprehensive picture showing that caesium continues to be a big problem in many countries.

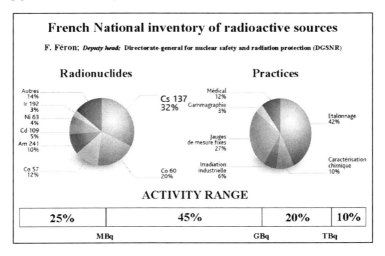

The developing countries also participated and we were presented with a data of inventories from India and when you look at the table below you will see that this data is not complete.

Major activities involving Radioactive Sources in India

Management of Radioactive Sources Ensuring Safety and Security : The Indian Scenario; J.K. Ghosh, Board of Radiation & Isotope Technology; Mumbai, INDIA

Devices	Sources	Number
Telegamma units	^{60}Co (+ depleted uranium used as shielding material in certain cases)	272
Brachytherapy units	^{60}Co, ^{192}Ir, ^{137}Cs, ^{90}Sr	229
Gamma irradiators	^{60}Co	12
Gamma chambers	^{60}Co	110
Industrial gamma exposure devices	^{192}Ir, ^{60}Co (+ depleted uranium used as shielding material in certain cases)	1182
Nucleonic Gauges including well logging sources	^{241}Am, ^{241}Am-Be, ^{137}Cs, ^{60}Co	7072
Medical and Industrial LINACs	Depleted uranium used as shielding material	64

This is just the materials that the Indians have under control – it is not an inventory of all the sources they have.

Excess Sources and Recoveries

One important issue that was raised was the excess sources problem. In the US in particular there are a very large number of excess sources in all the states. There is also a lot of work being done on recoveries. Even in Brazil there is high activity in this area – as one would expect after the experience of Goiânia.

The NNSA presented the programme for recoveries in the US and there has been a large amount of resources recovered. The lessons learned on recoveries were:

- Dispose wherever possible.
- Use existing infrastructures, government and private, and develop and employ new technologies.
- The average cost of recovery was very low averaging about $3,000.

- Finally, and most important, the owners are desperate for recovery, so participation is high.

What was also important was that the recoveries took place under the Tripartite Initiative (Russia-USA-IAEA). This is an initiative that was the creation of two people who I am honoured to call friends: one of them is here today, Jack Caravelli, and the other is in Moscow, Mikhail Ryzhov. They created this extremely important initiative which has just been finalised. And really this is the biggest recovery programme of radioactive sources that we have done until now. These two people had tremendous vision to do this: I was present at the negotiations and both of them were very difficult to deal with but they reached an agreement. The outcome was presented by Mr. Agapov, head of the Department of nuclear and radiation safety of FAAE and Mr. McGinnis, Office of Global Radiological Threat Reduction National nuclear Security Administration US Department of Energy. I should remind you that the Tripartite Initiative has a three-part steering committee comprising MinAtom, DOE and IAEA which also provides the project management.

Unfortunately the Tripartite Initiative covers all countries of the Former Soviet Union except Russia. However it was very successful in these countries:

- Armenia
- Azerbaijan
- Belarus
- Estonia
- Kazakhstan
- Kyrgyzstan
- Latvia
- Moldova
- Tajikistan
- Ukraine
- Uzbekistan

There were some very very powerful sources in these countries they are now secure.

Examples of Secured Sources

Installation MRH-gamma-100

Gamma installation «LBM-gamma-1M»

Gamma-irradiation bench SPG-04-02

Irradiation containers of gamma-installation HOS

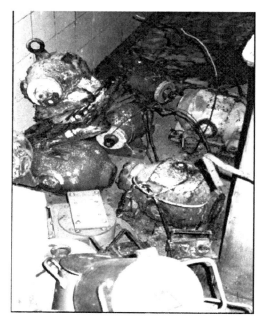

More important were the Gamma Kolos, sources of 3,500 Ci of caesium. All of them were secured – some in very bad condition. This was tremendous work as you can imagine – 3,500 Ci in each one of them.

Eight Gamma Kolos, in Kazakhstan, in 1974

Another big recovery operation was the Norwegian Russian co-operation involving the decommissioning of Radioisotope thermoelectric generators (RTGs) in north-west Russia. This was another enormous project – and was very successful.

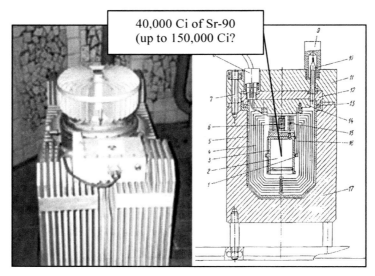

40,000 Ci of Sr-90 (up to 150,000 Ci?

Generators of the former USSR

These contain up to 40,000 Ci of strontium 90, and the larger ones up to 150,000 Ci. They were recovered in near Norwegian waters, transported by helicopter. They were taken from Murmansk to Moscow where they were dismantled and then sent to Mayak for recycling. I understand that this project has nearly been completed and not only are the Thermo generators recovered, and have been replaced by solar cell panels in those sites that are still operating.

Illicit Movement

In spite of all these recoveries, there continues to be illicit movement. Below is a table of confirmed incidents from 1993 to 2004 taken from the IAEA database – I believe that this database is about as reliable as the database the police have on drug movement: basically they know about approximately 1% of the drugs consumed in one big city and our database is probably as accurate. Nevertheless it gives us a picture.

If you look at the number of incidents by type from 1993 to 2004, you will see the absolute dominance of radioactive sources from nuclear materials.

Confirmed incidents 1993-2004

Illicit Trafficking involving Radioactive Sources, Richard Hoskins, **IAEA**

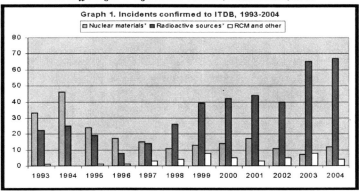

Also you will see the absolute dominance of caesium. Caesium continues to be a problem – too much caesium was produced.

Radioactive sources by type (1993-2004)

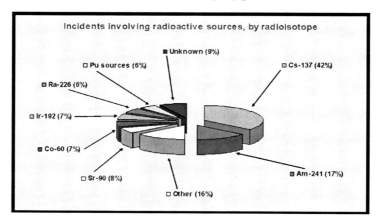

In analysing the categories of sources, in Categories 1 and 2 there are really very small amounts crossing borders and all are small sources in general.

Radioactive sources by category (1993-2004)

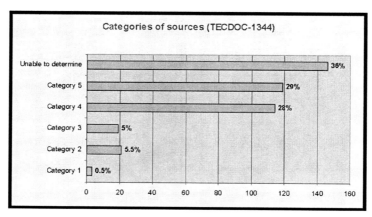

When it comes to recovery, this is very big and is increasing over time.

Recovered vs. Missing (1993-2004)

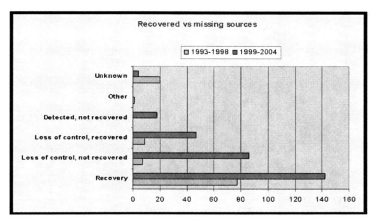

Another important issue is that criminal activity is not so dominant – the majority of cases do not involve criminal activity. They are Goiânia type activity – equally dangerous, but not criminal.

Criminal Activity (1993-2004)

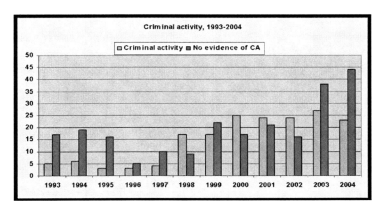

Another important thing is that the traffic is detected at borders. This confirms a paper presented by the Russian Customs which stated that detection by intelligence/CIA was very nice but didn't work. The real detection is at the borders.

Cross Border Movements (1993-2004)

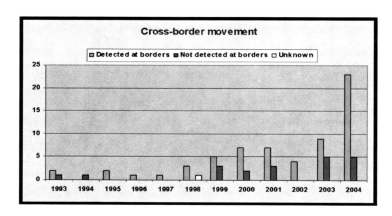

The conclusion that we can draw from all this:

- 'Only' 11% of sources involved in incidents were dangerous. But they are increasing.

- Less than half of incidents show evidence of criminality and many were amateurish, supply-driven and opportunistic. But well-organized criminals are less likely to be detected. There are indications of improved accounting methods and increasing effectiveness in radiation detection at borders and elsewhere.

- Increasing number of incidents involving radioactive sources.

- Open and other sources show a large number of cases as yet unconfirmed.

- Far too many sources are lost or stolen without being detected or recovered.

- We don't know what we don't know.

But really the most important of these is the last one "we don't know what we don't know".

Border Control

There has been an immense improvement in the control all over the world. First of all there are training exercises between countries – this is something new. A Turkish Bulgarian exercise is just one example.

A very interesting paper from Nikolay Kravchenko, Vice-Chief, Directorate of Special Techniques and Automation of Technologies, Russian Customs, identifies two things we didn't know: first, that these systems are triggered many times but only a few are in response to illicit trafficking.

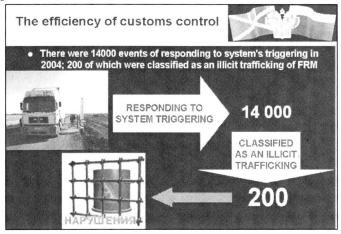

And second, that 95% of incidents are discovered by monitoring, not by documentation or intelligence.

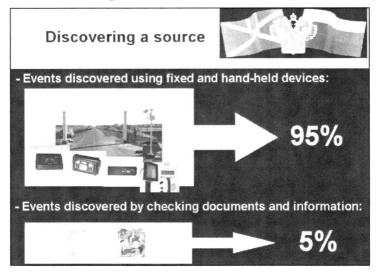

Another thing I should like to add is that border monitoring takes place in, shall we say, unexpected countries – Syria for example. And even in countries like Madagascar, where years ago not many people even knew the word radioactivity, now they have very sophisticated equipment for border control.

Implementation of the IAEA Code of Conduct

In the implementation of a code of conduct we are not doing very well. There is a very big commitment to it, but the reality is that there are many doubts on the actual implementation. There is no follow up mechanism. There is no international appraisal in place to see whether or not the code is being implemented.

Nevertheless there are interesting examples of where it is working. For example in Belarus there are a big problems regarding the borehole repositories used in the past by the USSR military units. No documents relating to the design or the radioisotope content are available, and in many cases the repositories are located in desolate areas where appropriate security cannot be provided.

China presented for the first time the big organisational problem which they have. There is a lot of work to be done in China as you can imagine. It is a big country with absolutely no control in places and with thousands and thousands of sources.

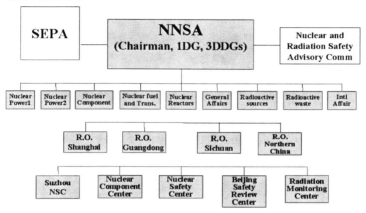

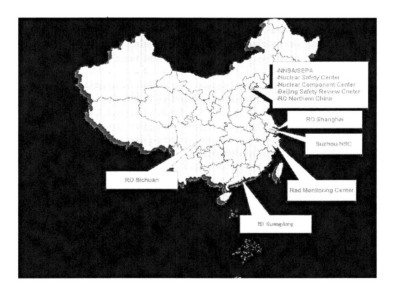

In Europe things should be simpler. Thanks to the Treaty of Rome, everything is concentrated in Brussels. However up till now they only had this directive 96/29 that relates to the implementation of the code.

Europe

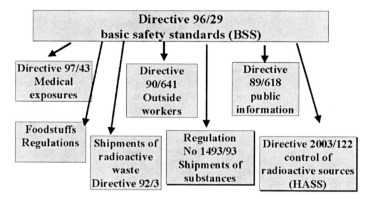

In the US, and this is something we have discussed many times with the Nuclear Regulatory Commission, there are very small differences in the definitions of radioactive sources and this causes confusion in some other countries.

High Risk Radioactive Material

Radioactive Material	Category 1		Category 2	
	Terabequerels (TBq)	Curies (Ci)	Terabequerels (TBq)	Curies (Ci)
Americium-241:...............	60	1,600	.6	16
Americium-241/Be:..........	60	1,600	.6	16
Californium-252:..............	20	540	.2	5.4
Curium-244:....................	50	1,400	.5	14
Cobalt-60:.......................	30	810	.3	8.1
Cesium-137:	100	2,700	1	27
Gadolinium-153:..............	1,000	27,000	10.0	270
Iridium-192:....................	80	2,200	.8	22
Plutonium-238:...............	60	1,600	.6	16
Plutonium-239/Be1:.........	60	1,600	.6	16
Promethium-147:.............	40,000	1,100,000	400.0	11,000
Selenium-75:...................	200	5,400	2.0	54
Strontium-90:..................	1,000	27,000	10.0	270
Thulium-170:...................	20,000	540,000	200.0	5,400
Ytterbium-169:................	300	8,100	3.0	81

One important thing to note is the prioritisation for compliance of sources in the USA.

- High priority – Panoramic Irradiators; Manufacturers/Distributors

- Medium priority – Medical and Research facilities, Radiography, Well Logging, Broad-scope licenses, self-shielded irradiators, open-field irradiators, and other licensees.

- Low priority – Fixed gauges. These are still being discussed for example of the DoE auditor thinks that fixed gauges should not be low priority because many can be added together which would therefore increase their priority.

Another interesting exercise is to examine the progress in countries like Nigeria where they are already providing authorisations for import and export of radioactive materials – Africa was one of the regions that used to be completely out of control.

Authorizations granted for import and export in 2004 and the 1q 2005

Year	Import	Export
2004	46	19
1Q2005	11	1
Total	57	20

Import, export and maximum activity of sources for 2004

Sources	Number Imported	Number Exported	Max. Activity (Ci) Import	Export
Ir-192	63	44	5239.56	875.18
Cs-137	4	2	0.59	1.97
Se-75	7	14	747.20	369.50
Total	74	60	5987.35	1246.65

Japan still is a problem, but they have their own solution – the real control in Japan is done through industry, the Japan Radioisotope Association more than through regulators. The real control in Japan is the industry not the regulators. I will make no judgement about this but in the rest of the world we have kept industry outside – in Japan it is just the opposite, the control is done through industry.

Distribution and Transfer of Sealed Radioactive Sources in Japan

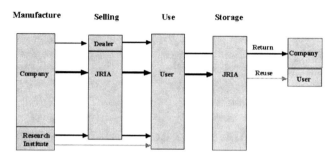

JRIA : Japan Radioisotope Association

Another area of enormous success is the monitoring of scrap metal. For example in Turkey, where there used to be numerous accidents.

And in Spain where there was an accident near Gibraltar where scrap was melted and pollution was detected as far away as Germany. As a result they have the monitors nearly everywhere covering all types of sources.

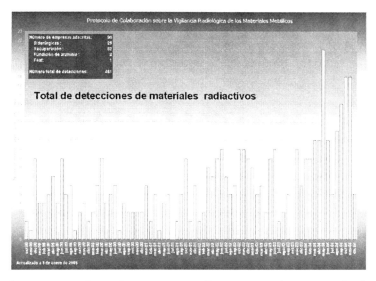

And we can see how the recovery of sources or the removal of sources has increased.

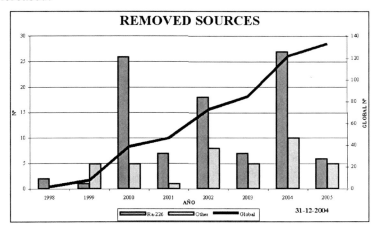

One interesting factor is the number of sources of radium in scrap metal – and radium can be a very big problem, not just caesium.

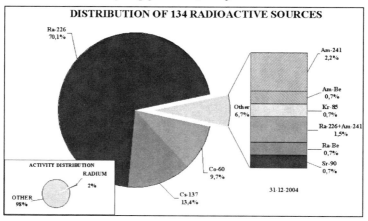

Disposals

This is a problem we have not solved, and we are still playing with the borehole disposal concept. South Africa has been working very hard on this. Not very different to the old boreholes that were used in the Soviet Union but a little more sophisticated.

AFRA Borehole Disposal Concept

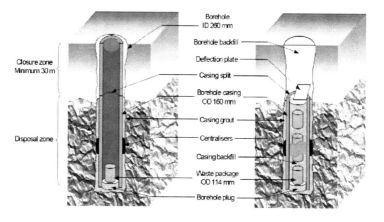

Bilateral Cooperation

This has increased enormously. The Australians were example have lodged a programme to cover the non-member states of the IAEA in the region. And remember that radioactive sources are everywhere.

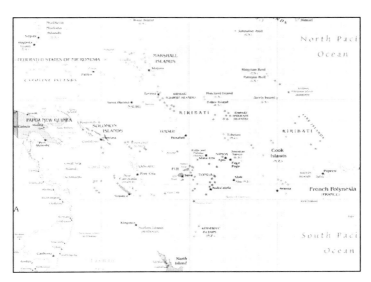

Japan has enormous bilateral cooperation in the Philippines where they have been doing radium conditioning and collection of Am 241 sources and a national monitoring system for metal products. In Thailand, where there was a cobalt 60 accident in 2001, Japan are now providing gate monitors at a steel company. In Indonesia they have provided a lot of security improvements. And in Korea they have helped with the regulatory system which was in a shambles.

International Services

There are also a lot of international services provided by the IAEA, one of which is the database held by the agency with all the source details, covering over 6,500 entries of sources.

There is also a database of all devices with over 7,900 entries.

And a manufacturer and distributor database with over 1,260 entries.

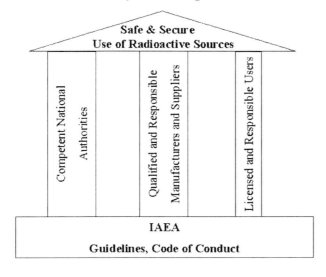

Another important service is the regulatory authority information system and this is now being used by 214 participants from 99 countries.

Industry

We had been keeping industry out and there was a paper from the industry outlining how they see their participation in implementing the code of conduct.

A Source Life Cycle Management Structure

What Do We Do If Something Happens?

There was a presentation from New York which has a lot of experience in terrorist attacks. But frankly speaking I was very surprised that, having a very big and complicated organisation, some basic elements were not there. For example, if a fire-fighter asks "Can I go into that area or not?" there was not always an answer – all the instrumentation they had to work with was too sophisticated. And this was in a city that has probably the biggest terrorist experience in the world.

Leonid Bolshov, of the nuclear safety Institute (IBRAE) of the Russian Academy of Sciences radiation exposure in the event of a radiological attack. The basic information is now available on their website, although you will need to buy the publication if you want to know the full details. It covers many things, but the most important is the protection of rescuers. You will remember that we have in the standards norms for normal situations, but using those numbers will not solve a problem. There is now agreement on the numbers for rescuers and these numbers should be the basis for simple detectors that fire-fighters should carry with them.

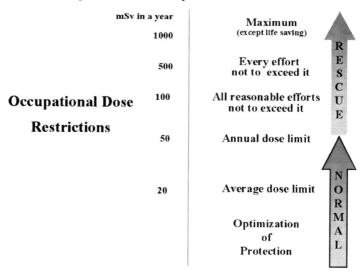

Also there was a very difficult agreement covering female rescuers – female rescuers cannot be permitted to operate in rescue situations, because it is absolutely impossible in these situations to protect the unborn child and infants, if she is feeding an infant. This will be very difficult to implement, and there will be a lot of reaction to it, but it is a scientific reality.

Protecting the Public

You probably know that the current international radiation protection approach has two standards. One for new activities, and one for existing situations. For new activities the dose limits are very, very stringent. And probably the public will ask for these limits to be applied. But frankly speaking can these levels be applied in practice? There is a dirty bomb and there is an increase in radiation and we had to control this increase. The system was not designed for this. What is recommended is that it be treated as an intervention after the attack. There are two questions that need answers. Should the resulting contamination be reduced? If the answer is yes, by how much? These are the two questions. The criteria for intervening shown below.

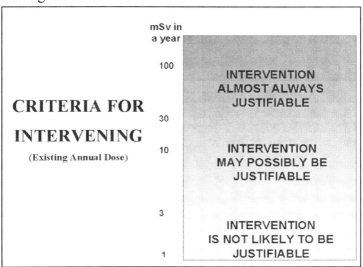

Also there are recommendations for the avertable effective doses, for sheltering, for temporary evacuation, for relocation and for iodine prophylaxis.

Avertable effective doses
that would generically optimize specific countermeasures

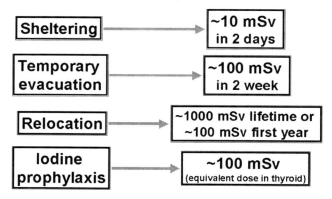

Dealing with the Aftermath

This is a very complicated problem. Regarding the safety of radioactive waste management, Goiania provides a benchmark: 1,000 Ci will give approximately 5,000 m^3 of waste. And the IRCP publication gives some indication of the numbers for dealing with long-lived radioactive residues, but the return to normality will be very difficult. The termination of protective actions will be plagued by fears of long-term effects – and the problem will be the media.

Professor Bolshov has data on how people perceive the effects of radiation. In Hiroshima the 'late' deaths were 421 people, yet a poll of supposedly intelligent people said they were 750,000. It was the same for Chernobyl – actually between 60 and 80 'late' deaths, recently reconfirmed, yet people believe that thousands of people died.

	Actual number of victims	Estimates made by the students of the natural science departments (average)
Hiroshima	Instant and short-term deaths 210 000	270 000
	Late deaths 421 (from 86572) hibakushi	750 000
	Early deaths 31	40 000
Chernobyl	Late deaths ≈ 60 – 80 (liquidators and children in the Bryanks region)	253 000

Perception: how many victims from Hiroshima and Chernobyl?

The problem that we have is that the effects of radiation cannot be detected unless you screen a lot of people.

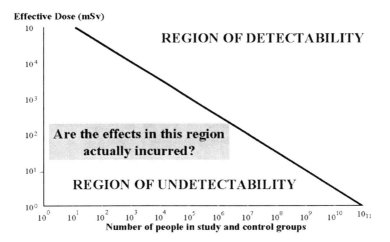

Are the effects in the region of unpredictability actually incurred? We don't know. There is an epistemological limitation to our knowledge. We have no basis for our knowledge, and this will be very difficult to convey to people.

Outlook

There continued to be shortfalls in the security of radioactive sources.

- Implementation at the national level of the code of conduct
- The guidelines for export import
- The recovery of orphan sources
- The strengthening of national controls
- And the implementation of model projects in non-member states is still very low

There are still a lot of things to be done to ensure implementation at the national level.

The Way Forward – International Co-operation

I have always said that the way forward is through international cooperation. And NATO could do a lot to help this. As you know NATO is working in this area with the NATO Security Through Science programme and the NATO Weapons of Mass Destruction centre. The latter is a new centre whose objectives are:

- To ensure a vigorous debate in NATO leading to strengthened common understanding among Allies on WMD issues and how to respond to them
- To improve the quality and quantity of intelligence and information sharing among Allies on proliferation issues
- To support the development of a public information strategy by Allies to increase awareness of proliferation issues and Allies' efforts to support non-proliferation efforts
- To enhance military readiness to operate in a WMD environment and to counter WMD threat
- To strengthen the exchange of information concerning national programmes for bilateral WMD destruction and assistance
- To enhance the possibilities for Allies to assist one another in the protection of their civil populations against WMD risks

The only problem that I see in these objectives is the word "Allies". Because NATO can have a role that extends beyond just Allies – it can help other people as well. Whether that is possible or not, I have no idea. But if the role of NATO is only limited to Allies, it is not enough – because the big problems are elsewhere.

The IAEA

The IAEA can continue to do a lot particularly since it received the Nobel Peace Prize in 2005. And it got the Nobel peace prize not only for its efforts to prevent the use of nuclear weapons for the military, but to ensure that nuclear energy for peaceful purposes is used in the safest possible way.

After the award of the Nobel Peace Prize, the media only emphasised the verification role of the IAEA – it paid no attention to its safety and security role. Yet the Nobel Prize was awarded for both. The IAEA is the only UN agency with specific statutory responsibilities on safety and security. We have to profit from that: it has three simple functions: to establish standards; to provide for their application; and to facilitate international conventions.

Continuing Issues

The issues regarding the Security of Radioactive Sources continue to be simple to formulate and to resolve. They are:

- The prevalence of sources
- The orphanage issue
- Relinquished control
- The isolated state of manufacturers – except in Japan
- The problem of unconventional sources

My solutions are simple:

- **For the prevalence issue**: since radioactive sources are abundant and widespread all around the world, the solution is to internationalise control.
- **For the orphanage issue**: since many radioactive sources go astray, the solution is to help find them and help regain control of them.

- **For the relinquished control issue**: since control of radioactive sources is relaxed, even those that are well regulated are not especially well secured, the solution is to impose international prescriptive regulatory requirements for ensuring control (commensurate with the risk).

- **For the isolated state of manufacturers**: the solution is to recognize and promote the recently created Association of Manufacturers.

- **For the problem of unconventional sources**: many orphan sources are special and powerful, the solution is to exert international pressure on countries that produced and abandoned these sources.

Immediate Actions

There are two immediate actions that need to be undertaken. First to profit from the legacy of past activities – the Tripartite experience should be absorbed and extended to other parts of the world. And we should launch other initiatives. But no international initiatives can replace a country's own actions. The time is right for international binding obligations to recover orphan sources.

For the sustainability and continuity of control, the IAEA Assistance experience should be absorbed and others should be launched – but again no international initiatives can replace a country's own actions. The time is right for international binding obligations to ensure that the provisions of the Code of Conduct are followed by all.

To summarise: the time is right for binding commitments for a harmonized, effective and sustainable international regime of the safety and security of radioactive sources, because there are many countries with problems. Each one of the member states of the United Nations has problems. I now believe that we should go back to the Dijon proposals and insist that our political masters should work towards an international convention on the safety and security of radioactive sources.

Because there will be no security for any of us unless there is security for all of us. In my opinion we are confronted with a difficult dilemma between overreaction and responsibility.

Finally, let me add a small advertisement: there will be an IRPA congress in Buenos Aires in 2008 in which one of the subjects will be the Security of Radioactive Sources. More than 2,000 people will be there and for the first time the Security of Radioactive Sources will be a prominent agenda item.

SECTION III

HIGH CONSEQUENCE RADIOACTIVE TERRORISM SCENARIOS

1 HIGH-CONSEQUENCE RADIOACTIVE TERRORISM SCENARIOS – RUSSIA

Dr. Leonid Bolshov
Director, Nuclear Safety Institute (IBRAE), Russian Academy of Sciences, Russian Federation

NATU Advanced Research Workshop
International Approaches to Securing
Radioactive Sources Against Terrorism
London, 04-06 November, 2005

Round Table
"HIGH CONSEQUENCE RADIOACTIVE TERRORISM SCENARIOS"

Leonid Bolshov
Nuclear Safety Institute (IBRAE) Russian Academy of Sciences
www.ibrae.ac.ru

Almost every week in Russia there is news of another terrorist event. In the professional community we are waiting for news that an event could be radiological in nature. We have had the Chernobyl experience and we understand very well that prevention and preparedness are key.

IRS: safety and security problems

2

- IRS represent an outstanding consequence from the development of nuclear technology;
- IRS are widely distributed and effectively used all over the world;
- In the decades of IRS application, certain problems have appeared in providing their safety and security;
- The potential threat of radiation terrorism has resulted in new challenges in the field of IRS safety.

Of the potential tools for radiation terrorism SNF, RAO, and weapons material IRS are the most effective. Mankind did indeed develop a very good technology.

Application of Radionuclide Sources

3

Equipment containing radionuclide sources is widely used in different industries, namely:

- Nuclear Power and Engineering,
- Research,
- Metallurgy,
- Geology,
- Mining,
- Meteorology,
- Chemical and Petroleum Industries,
- Medicine and Agriculture.

A small amount of RA material is very effective at producing a lot of radiation and thus has numerous applications worldwide. This widespread use has led to losses, misuse and so on. And the increased threat of terrorism adds new challenges. There are a multitude of diverse methods:

Mechanisms of Dispersion in the Environment

4

- blasting (aerosols, gases);
- thermal effect (aerosols, gases);
- dispersion of liquids (aerosols, vapour, steam);
- dilution in aquatic environment;
- installation of IRS in public places
- and on, and on, and on.

And there is a limitless range of targets:

Targets for radioactive contamination

5

- big cities;
- public places;
- critical infrastructures;
- transport communications;
- drinking water supply;
- foodstuff, pharmaceuticals, etc.;
- and on, and on, and on.

The problem is not theoretical – it has already occurred.

Attempts of the use of radioactive substances for terrorist missions

6

Publicized in mass media

- November 1995 Chechen militants threatened to detonate containers with radioactive cesium in the Izmailovo park in Moscow.

- May 2002, special services of the US detained the person suspected in preparation of the act of terrorism with a "dirty bomb". That was Al-Mudjahir, who is known as Jose Padilla, arrived from Pakistan. According to special services information, he was preparing to explode RDD in order to radioactively contaminate the USA capital.

- 2004, UK citizens, who were trying to create a so called "dirty bomb" and explode it in London, were detained being suspected of terrorism.

And the range of sources in the public and private sectors is widespread and tempting.

Application of Radionuclide Sources

7

- Radionuclide sources for electrical supply (radioisotope thermo electrical generators - RITEG) used as autonomous power supply sources in remote regions and regions, which are difficult of access (radio- and light beacons, meteorological posts, etc.);

- Radiation-technological installations (RTI) for medical material sterilization and treatment of agricultural products, industrial and domestic wastes;

- Apparatuses for radiotherapy (treatment of cancerous growths, cancers of breasts, lungs, gullet, mouth, larynx, urinary bladder, etc.);

- Devices for non-destructive examination (gamma-defectoscopes);

- Equipment for process control (densitometers, level gages, thickness indicators);

- Composition analysers.

Ionizing Radiation Sources manufactured in USSR

8

No.	RITEG type	Initial rated thermal power W	Initial rated activity, Ci	Output rated electrical power, W	Output voltage, V	Weight, kg	Start of production
1	Beta-M	230	35000			560	1978
2	Efir-MA	720	111000	30	35	1250	1976
3	RITEG IEU-1	2200	339000	80	24	2500	1976
4	RITEG IEU-1M	2200 or 3300	340000 or 510000	120 or 180	28	2 or 3 items, 1050 each	1990
5	RITEG IEU-2	580	89000	14	6	600	1977
6	RITEG IEU-2M	690	106000	20	14	600	1985
7	Gong	315	49000	18	14	600	1983
8	Gorn	1100	170000	60	7 or 14 or 28	1050	1983
9	Senostav	1870	288000			1250	1989

^{137}Cs-based source

9

Nuclide	Capsule construction materials	Source activity	Critical temperatures
137Cs	Cs is incorporated in glass ceramics. Glass ceramic tablets are encapsulated in stainless steel	up to 185 GBq	Melting temperature of glass ceramics exceeds 1700°C. While melting, temperature of stainless steel is below to 1700°C

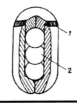

1 - capsule
2 - active part

10

Industrial sources

Nuclide	Capsule construction materials	Source activity	Critical temperatures
^{85}Kr	Two-cavity source is made of titanium. The internal cavity contains ^{85}Kr	37 GBq	Melting temperature of titanium is 1668 °C
^{241}Am	Sealed stainless steel capsule. Target window is made of 0.1 mm foil	$6*10^6$ – $3*10^8$ Bq	The target window can easily be mechanically damaged. Melting temperature of stainless steel (12X18H10T) is up to 1700 °C
^{60}Co	Stainless steel (12X18H10T)	1500 – 3700 GBq	Melting temperature of stainless steel (12X18H10T) is up to 1700 °C

11

RITEG <Gorn>

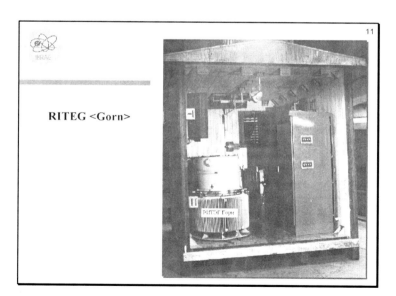

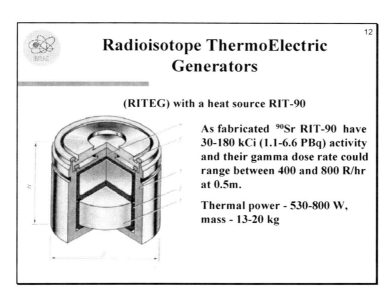

The International Community is moving, but too slowly to deal with the threat.

It is not a surprise that IAEA and other national and international organizations increased their efforts to secure IRS last time.

Dynamics of radiologically hazardous incidents

14

With ionizing radiation sources at Russian facilities (other than RF Minatom's) in 1997-2001

Incident type	1997	1998	1999	2000	2001
Source failure (depressurization)	8	3	6	10	17
Thefts of sources	13	22	3	6	6
Discovery of non-accounted sources	14	16	5	1	2
Breaks of shells with a source in wells while conducting geophysical survey	9	10	14	18	24
Losses of sources while transporting	-	5	1	2	1
Deliberate depressurization of sources	-	2	-	-	-

Some examples

15

- According to GAN of Russia <u>five</u> incidents related to the breakdown of IRS-containing well tools during geological works were detected in the IV quarter of 2004.
- In most similar cases removal of radioactive source from the well can not be managed.
- Such incidents may potentially result in the fact that radioactive substances will fall in the terrorists' hands.

16

- According to GAN of Russia <u>four</u> IRS thefts were detected in 2003 in Russia.

- Among them is a theft, which was detected in March 2003, of Beta-M RITEG belonging to one of the military units. This source without any biological protection was found at the coastal line under the ice. Dose rate above the water surface exceeded 30 R/h.

- The source was removed from the water, placed into the container and accepted for storage.

According to State Sanitary Regulatory Body 4 high-power IRS were detected in scrap metal in 2002 in Primorsk region.

The last incident of such nature took place on June 22, 2005, in Vladivostok. Sr-90 source was been found in the port in scrap metal to be shipped to China. It was previously used in a density measurement tool. Its dose rate was 300 мкR/h.

18

Other countries are in no better position

"... in the United States alone,
375 radioactive sources on average
are lost or stolen annually".

(John Bolton, Report at the IAEA conference on Nuclear and Radiation Safety,
September 30, 2002, London)

19

Direct Impact of Radiation

- Acute exposure of people to significant doses of radiation in a short time period (hours or days) can cause acute radiation syndrome with heavy damage to the health or threatening the life of the victim.

- Protracted radiation exposure of people owing to significant contamination of the environment by radioactive materials, might cause extended consequences (from months to dozens of years) — with an increase in morbidity and mortality and, in particular, of oncological sickness.

Of course, dispersion of one big source 10^3–10^4 Ci within a big city will result in severe acute or protracted exposure. This would involve the relocation of millions, and the loss of thousands of lives.

20

Indirect Impact of Radiation

Social, economical, political, psychological, and demographic consequences:

- Economical damage owing to expensive necessary countermeasures (evacuation, ecological remediation of large areas, compensation for lost or devalued property, etc.)

- Malfunctions in the infrastructure and commercial activity owing to significant radioactive pollution, even if there is no actual hazard for human health.

21

Indirect Impact of Radiation (cont)

Social, economical, political, psychological, and demographic consequences (cont):

- Worsening of demographic situation (abandoned delivery, migration of population from the contaminated territory) related to the overestimated perception of the radiation hazard.

- Increased tension in social relations owing to imperfections in radiation protection legislation and lack of universally recognized sources of reliable, trustworthy information about actual the radiation situation and its associated risks - resulting in various interpretations of the situation and rumors which usually drastically overestimate the actual hazards.

It is less obvious that even a small or limited activity dispersal in a highly populated area of a big city could cause an immediate indirect impact on most areas of human life.

Economic Damage Caused by Radiation Terrorism Event (RTE)   22

Direct radiation impact*

- Humanitarian damage (lost lives and health) and related economic damage (compensation, medical expenses)

- Lost industrial and commercial facilities

- Expenses for mandatory countermeasures (evacuation, ecological remediation of large area), including unjustified intervention

* these damages are known as "direct costs"

Economic Damage Caused by Radiation Terrorism Event (RTE)   23

Indirect radiation impact

- Reduced output in various sectors of the economy*
- Broken social, economical, and industrial connections
- Depreciated real estate
- Decrease in commercial activity in the contaminated zone and neighboring territory (owing to broken business connections)
- Lost profits from tourism, sport and culture events etc.
- Compensations for depreciated property
- Hidden losses caused by increased negative attitude of the community to radiation and, in particular, to the nuclear power industry

* often called "indirect costs"

Economic damage caused by RTE is rather difficult to quantify in detail. However our experience of Chernobyl has taught us a lot.

24

Highly sensitive perception of radiation risks by the population and imperfect normative and legal basis in the field of radiation protection of population and environment could result in significant increase in damage

Public perception is the real explosive in an RTE event.

25

Perceived Increase in the Scale of the Accident.
Ratio lost:victims:exposed:involved:alarmed in the case of
a **Zero** Radiation Accident

**Balakovo NPP,
November 4–6, 2004**

- Emergency training and triggering of the emergency protection system (detector «steaming»)
- Event of the level «**0**» in INES scale
- Late information from officials
- Panic spreading: approx. 25m persons from 11 regions of Russia for 30 hours
- Iodine poisoning, excessive spirit consumption

*This picture
appeared in all
Russian mass media
for 3 days*

Sometimes response is very intense, even if the real impact is zero. For instance, Iodine poisoning was experienced by self-administration of medicinal iodine by individuals who thought they were at risk.

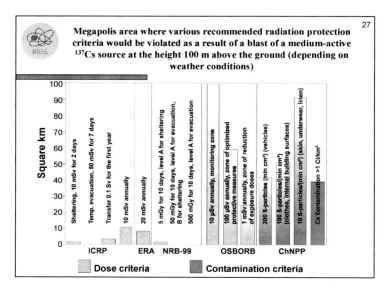

Megapolis area where various recommended radiation protection criteria would be violated as a result of a blast of a medium-active ^{137}Cs source at the height 100 m above the ground (depending on weather conditions)

Norms and recommendations what to do in a case are so unclear and sometimes controversial. For decision makers it would be a mess to select which one to use. Our Chernobyl experience and experiences of response to real events tell us that the lowest possible choice would not be a surprise.

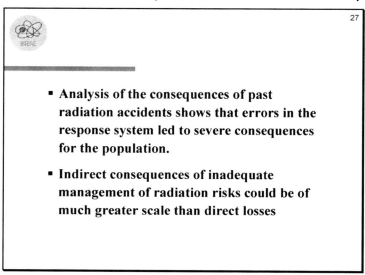

- Analysis of the consequences of past radiation accidents shows that errors in the response system led to severe consequences for the population.

- Indirect consequences of inadequate management of radiation risks could be of much greater scale than direct losses

Any errors in the emergency response system will result in severe consequences for population.

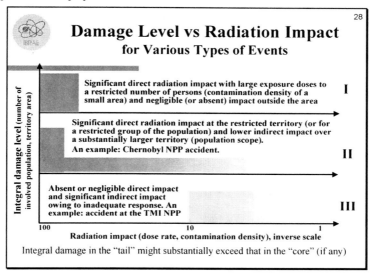

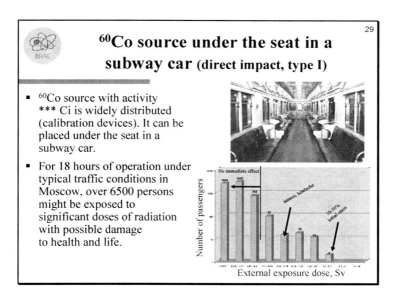

Co-60 source with activity 20 Ci (740 GBq)

30

Blasting of ^{90}Sr IRS at a subway platform during rush-hour
(direct and indirect impact, type II)

- Widely distributed industrial source ^{90}Sr with activity *** Ci is blasted at the subway platform during rush-hour, when over 1200 persons can simultaneously be present at the station.

- Direct impact: during the first 15 minutes following the accident, over 300 persons would be exposed to hazardous radiation doses owing to radionuclide inhalation (over 5 Sv lung exposure). 500 additional persons would obtain ~1.5 Sv of lung exposure, which is also above the threshold of radiation-induced lung sickness.

Exposure dose, Sv	Zone 1 (320 pers.)*	Zone 2 (480 pers.)	Zone 3 (480 pers.)
Skin	0.36	0.08	0.02
Lung	5.8	1.4	0.42
Whole body (effective)	0.72	0.16	0.04

* depending on the distance from the source

Sr-90 source with activity 2 Ci (74 GBq)

31

Blasting of ^{90}Sr IRS at a subway platform during rush-hour (cont'd)

Indirect impact:

- the contaminated station cannot be used by passengers for entering, leaving, and interchange. It results (at least, in Moscow), in heavy traffic conditions for a long time period (years). Simultaneous contamination of several stations and the transfer paths would virtually paralyze the operation of the subway as a whole;

- vast medical examination (with respect to involved population, cost, and labor expenses) followed by protracted monitoring of the suffered persons would be required.

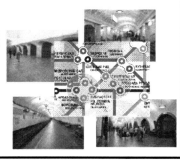

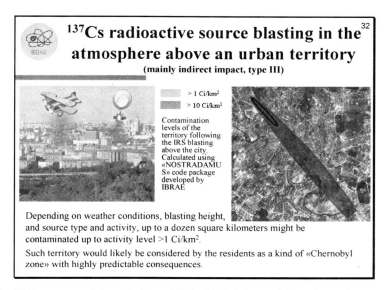

Cs-137 source with activity ~1000 Ci (37 TBq) blasted at 100 m above the surface.

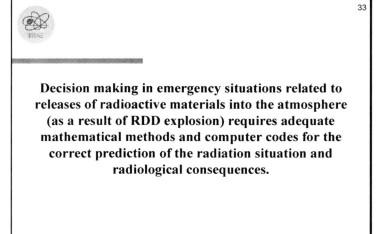

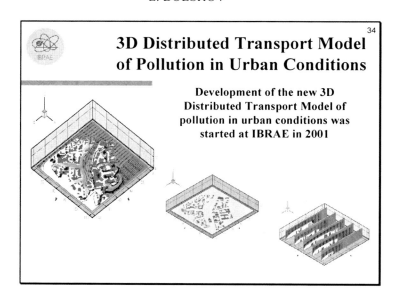

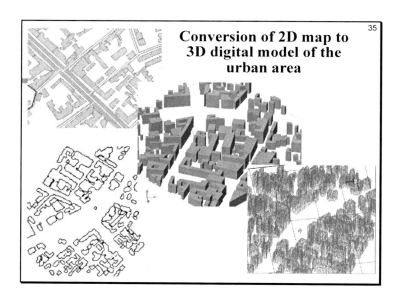

36

The input data for calculations

Radioactive filling of the RDD
- *Am-241 source used in oil well survey;*
- *Activity of source - ***;*
- *Power of blasting - 10 kg TNT;*
- *Initial height of the radioactive cloud - 20 m.*

Weather conditions
- *neutral atmospheric stability;*
- *wind speed (10 m) - 5 m/s;*
- *Calculation zone - 1 sq.km;*
- *Population density – 10,000 person/sq.km;*
- *In the blast period - 50% of people are inside and 50% are outside the buildings.*

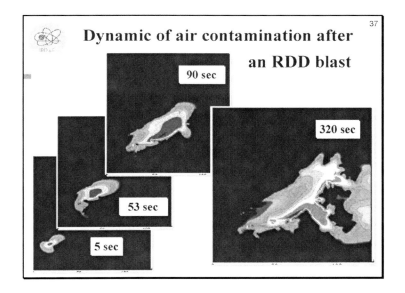

37

Dynamic of air contamination after an RDD blast

90 sec

320 sec

53 sec

5 sec

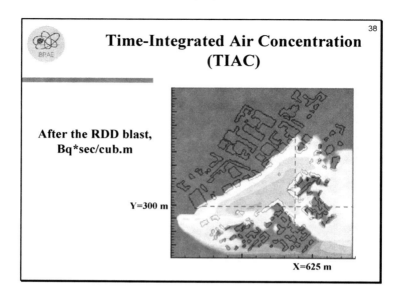

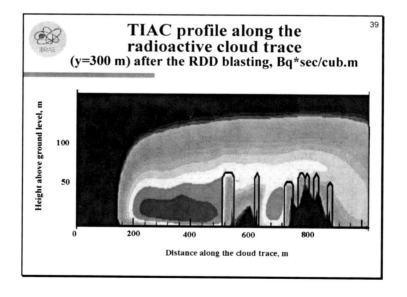

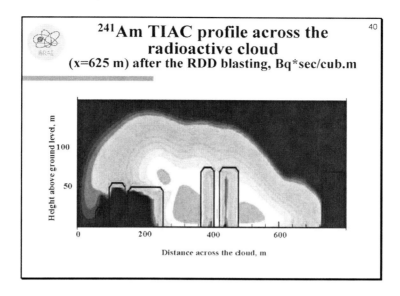

^{241}Am TIAC profile across the radioactive cloud
(x=625 m) after the RDD blasting, Bq*sec/cub.m

40

Predicted distribution of the population with respect to lung absorbed dose following a ^{241}Am «dirty bomb» blasting (direct impact)

41

Calculated using the 3D distributed transport model developed in IBRAE

Lung dose	outdoors	indoors
< 0.05 mSv	622	82
0.05-0.5 mSv	150	23
0.5-5 mSv	171	21
5-50 mSv	283	17
50-500 mSv	249	13
0.5-5 Sv	464	1
5-50 Sv	86	0
> 50 Sv	0	0
total	2023 (1082)	158 (31)

Figures in parentheses stand for number of persons in zones of protective measures.
Assumes average population density 10 000 persons per 1 km^2 of urban territory.

Am-241 source with activity 2.7 Ci (100 GBq)

The population density substantially exceeds the average value during major events like concerts, performances, sport competitions etc. The potential number of RTE victims in this case might be very high

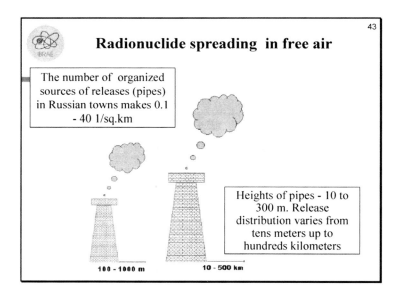

Radionuclide spreading in free air

The number of organized sources of releases (pipes) in Russian towns makes 0.1 - 40 1/sq.km

Heights of pipes - 10 to 300 m. Release distribution varies from tens meters up to hundreds kilometers

100 - 1000 m

10 - 500 km

44

**Road Pollution by Radioactive Materials
(indirect impact, type III)**

Radioactively contaminated water is spilled on the road pavement by a sprinkler or is spread along the road by vehicle wheels after spot spillage. As a result, road pavement over a large area (for Moscow, up to ~2 km^2) might be contaminated up to high level of several hundred Ci/km^2.

- **Direct impact** is virtually absent: significant exposure dose can be obtained only assuming protracted (several hours) stay at the contaminated spot (e.g. traffic-controller)

- **Indirect impact:** the contaminated territory should be deactivated, which requires complete removal of the pavement with traffic termination for the whole rehabilitation period. All operations should meet safety regulations, which makes them very expensive and time-consuming taking into account megapolis conditions.

Cs-137 source with activity up to 100 Ci (3.7 TBq)

45

Analysis shows that, from the point of view of radiological, social and economic consequences, a much broader range of sources and their results can be used as components of radiological weapons than is usually considered

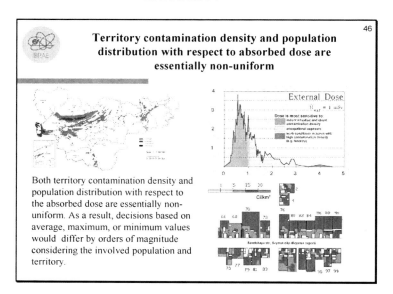

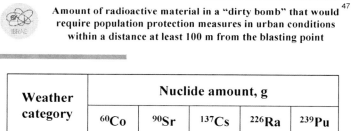

Key point – very small amount of wide-spread IRS material is considered enough in order to have a significant impact.

49

Potential hazards & severity of consequences

Factors determining the degree of potential hazard and severity of consequences of IRS application in terrorist attacks:

- Level of difficulty for creating RDD, means of their delivery and methods of dispersion;

- Direct and indirect damage to the population health, environment, social life and economics;

- Efficiency of national and international response systems on radiological terrorist attacks.

50

Potential hazards & severity of consequences

Factors determining the degree of potential hazard and severity of consequences of IRS application in terrorist attacks:
  ADEQUACY of:

- Informational, analytical and scientific base for solving the problems of prevention and minimisation of radiological terrorist act consequences;

- Normative and legislative base on radiation safety, protection of population and environment;

- Information of population on radiation risks.

CONCLUSION

There is an urgent requirement for national and international effort in order to:

- Improve IRS control and accounting systems
- Search for, control, and make safe spent, ownerless, and lost IRS
- Provide relevant control over IRS relocation
- Create an effective RTE response system
- Improve the existing legislative and regulatory system
- Develop and implement new approaches in radiation risk communications

2 HIGH CONSEQUENCE RADIOLOGICAL TERRORISM SCENARIOS – U.N.

Dr. Alex P. Schmid
Senior Crime Prevention and Criminal Justice Officer, Terrorism Prevention Branch, United Nations Office on Drugs and Crime (UNODC), Austria

UNITED NATIONS
Office on Drugs and Crime

High Consequence Radiological Terrorism Scenarios

Presentation prepared by Alex P. Schmid
(Senior Crime Prevention & Criminal Justice Officer,
Terrorism Prevention Branch, UNODC, Vienna) &
Robert Wesley (International Atomic Energy Agency, Vienna)

W.D. Wood and D.M. Robinson (eds.), *International Approaches to Securing Radioactive Sources Against Terrorism,*
© Springer Science+Business Media B.V. 2009

Four Types of Radiological Dispersion

1. A radiation dispersion device (RDD) in the form of a conventional explosive attached to radioactive material;

2. A silent dispersal of radioactive materials in the air, the water or on the soil, e.g. by aerosol, dilution or dusting;

3. A stationary radiological emission at a place where people reside (sometimes called a Radiological Emission Device),

and last but not least

4. An attack on a site containing radioactive materials and its dispersal there.

Types of Consequences of Radiological Attack

➤   Health impact on human beings exposed to radioactivity;

➤   Socio-psychological impact of the radioactive attack on direct and indirect victims, including changed behaviour of impacted segments of society;

➤   Political impact (including loss of confidence in government)

➤   Economic impact (caused by denial of access in contaminated zone on the one hand and costs of cleaning up the polluted areas on the other);

➤   Environmental impact.

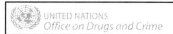

Effects of Whole-Body Exposure to Radiation

Low Consequences

Dose (REMS)	Short-term effects
0 -50	No detectable effects
➢ 50 – 100	Temporary depression of white blood cell count. Possible nausea at upper end of exposure range. No evacuation for temporary illness.

Source: Ken Larson. Nuclear Emergency. How to Protect Your Family From Radiation. Suwanee, Rhema Publishing, 1997, p.29 - 30.

Effects of Whole-Body Exposure to Radiation

Medium Consequences

Dose (REMS)	Short-term effects
➢ 100 – 200	Nausea, diarrhoea, possible loss of hair, temporary sterility, no immediately fatal but long-term risk of cancer
➢ 400 – 500	acute nausea and diarrhoea within a few hours. 50% of victims will die in about four weeks after showing signs of apparent recovery

Source: Ken Larson. Nuclear Emergency. How to Protect Your Family From Radiation. Suwanee, Rhema Publishing, 1997, p.29 - 30.

UNITED NATIONS
Office on Drugs and Crime

Effects of Whole-Body Exposure to Radiation

High Consequences

Dose (REMS)	Short-term effects
➢ 600 – 1000	Depressed white blood cell count, acute nausea and diarrhoea, 80% will die in four weeks after sometimes showing signs of recovery.
➢ 1000 - 5000	Virtually 100% chance of death, sometimes within a few hours

Source: Ken Larson. Nuclear Emergency. How to Protect Your Family From Radiation. Suwanee, Rhema Publishing, 1997, p.29 - 30.

UNITED NATIONS
Office on Drugs and Crime

Types of Consequences of Radiological Attack

- ➢ Health impact on human beings exposed to radioactivity;
- ➢ Socio-psychological impact of the radioactive attack on direct and indirect victims, including changed behaviour of impacted segments of society;
- ➢ Political impact (including loss of confidence in government);
- ➢ Economic impact (caused by denial of access in contaminated zone on the one hand and costs of cleaning up the polluted areas on the other);
- ➢ Environmental impact.

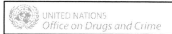

Three Levels of Socio-Psychological Impact of RDD Attack

1. Population in impact zone is evacuated in an efficient and orderly manner;

2. Authorities loose control over evacuation and uncontrolled mass exodus sets in but government regains control of the situation after some days;

3. Permanent regional population shift occurs as sources of employment around impact zone disappear due to the exodus of business and main social carriers of community life.

Three Levels of Political Impact of RDD Attack

1. Authorities cope effectively with RDD Attack and social cohesion is not affected

2. Authorities show incompetence and contradict each other, leading to widespread loss of faith in government

3. The impact of the event leads to a change in the political system (e.g. prolonged state of emergency)

Three Levels of Economic Impact of RDD Attack

1. Economic impact is local, temporary and decontamination costs are moderate;

2. Economic impact is national and prolonged and decontamination costs are substantial;

3. Economic impact is international and decontamination costs are very high.

Three Levels of Environmental Impact of RDD Attack

1. Environmental impact is local and of short duration

2. Environmental impact is local and of long duration

3. Environmental impact is regional and of long duration

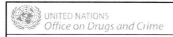

Types of Consequences	Level of Consequence		
	Low	**Medium**	**High**
Health	(Rems:0-100) Zero to temporary illness	(Rems:100-500) Immediate illness; up to 50% of victims will die	(Rems:600-500) Up to 80-100% fatality of affected population.
Psycho-Social	Evacuation is efficient and orderly/low psychological damage	Loss of control of evacuation but regain control in a few days/ psychological impact is apparent	Permanent population shift and serious business loss
Political	Affective government response and social cohesion not affected	Response displays incompetence/loss of faith in government	Change in political system
Economic	Local, temporary and decontamination costs are moderate	National, prolonged and decontamination costs are substantial	International and decontamination costs are very high
Environmental	Local and of short duration	Local and of long duration	Regional and of long duration

Commercially Available Radioisotopes which Pose the Greatest Security Risk

Element	Half-life
Americium 241	433 years
Californium 252	2.7 years
Cesium-137	30.1 years
Cobalt-60	5.3 years
Iridium-192	74 days
Plutonium 238	88 years
Radium-226	1600 years
Strontium-90	28.8 years

Source: Adapted from IAEA « Categorization of Radiation Sources"; cit. Charles D. Ferguson et al. Commercial Radioactive Sources : Surveying the Security Risks. Monterey, Center for Nonproliferation Studies, 2003, Table 5, p. 16.

UNITED NATIONS
Office on Drugs and Crime

High-Risk Sources

Practice or Application	Radioisotope	Radioactivity (Curies)
Thermoelectric generators	Strontium-90	20,000
	Plutonium-238	280
Sterilization and food irradiation	Cobalt-60	Up to 4,000,000
	Cesium-137	Up to 3,000,000
Self-contained & blood irradiators	Cobalt-60	2,400-25,000
	Cesium-137	7,000-15,000
Single-beam teletherapy	Cobalt-60	4,000
	Cesium-137	500
Multi-beam teletherapy	Cobalt-60	7,000
Industrial radiography	Cobalt-60	60
	Iridium-192	100

Source: IAEA. Categorization of Radioactive Sources. Vienna, IAEA-TECDOC,1344, July 2003

UNITED NATIONS
Office on Drugs and Crime

High-Risk Sources (cont.)

Practice or Application	Radioisotope	Radioactivity (Curies)
Calibration	Cobalt-60	20
	Cesium-137	60
	Americium-241	10
High/medium-dose brachytherapy	Cobalt-60	10
	Cesium-137	3
	Iridium-192	6
Well logging	Cesium-137	2
	Americium-141	20
Level and conveyor gauges	Cobalt-60	5
	Cesium-137	3-5

Source: IAEA. Categorization of Radioactive Sources. Vienna, IAEA-TECDOC,1344, July 2003

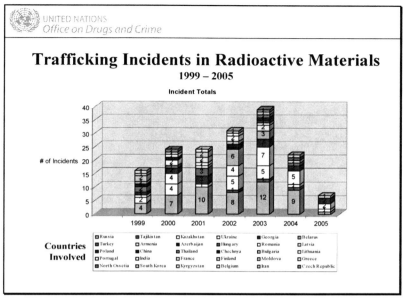

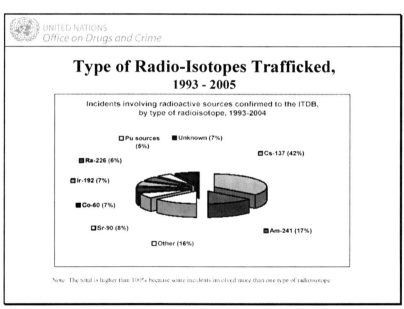

UNITED NATIONS
Office on Drugs and Crime

Sources of Trafficked Radioactive Materials
1993 - 2005

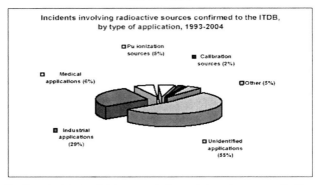

Incidents involving radioactive sources confirmed to the ITDB,
by type of application, 1993-2004

Note: The total is higher than 100% because some incidents involved more than one source with different applications.

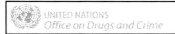

UNITED NATIONS
Office on Drugs and Crime

Tentative Conclusions from Trafficking Incidents
1999 - 2005

The Human Element:

➤ **Organization:**
  - ? Incidents show a continued mixture of organized and non-organized trafficking.
  - ? Organized groups are present and increasingly prevalent.

➤ **Technical Expertise:**
  - ? Significant number of incidents where technical expertise was present.

➤ **Perceived Demand:**
  - ? The perceived demand remains alarmingly high.
  - ? Perceived prices alarmingly high.
  - ? Perceived risks are still acceptable to traffickers.
  - ? The market remains active and is forecasted to remain so.

UNITED NATIONS
Office on Drugs and Crime

Tentative Conclusions from Trafficking Incidents
1999 - 2005
The Material Element

➤ **Volume of Incidents and Location:**
  - ? Still a rampant problem with 156 incidents in 51 countries.

➤ **Nuclear:**
  - ? On the basis of the number of known incidents of HEU and Pu,the quantities involved are insufficient for production of an IND.

➤ **Radiological:**
  - ? High volume of likely RDD materials.
  - ? Significant risk of terrorist groups procuring radioisotopes via illicit trafficking.

UNITED NATIONS
Office on Drugs and Crime

Consequences of the 1986 Chernobyl Nuclear Power Plant Accident

➤ 203 people were hospitalized during and after the ten days of reactor fire of whom 31 died (28 of them from acute radiation exposure)

➤ 116,000 people had to be evacuated following the accident

➤ An additional 210,000 people were resettled between 1990 and 1995 in Belarus, the Russian Federation and the Ukraine

➤ About 4,000 square miles of land were taken out of human use for an indefinite time; altogether more than 200 km2 in Europe were contaminated with cesium-137; 71% of which in three countries: Belarus, Russia and the Ukraine

➤ Between 300,000 and 600,000 people were involved in the cleanup of the 30 km evacuation zone around the reactor in the following months and years

➤ $ 12.8 billion cost of the disruption to the Soviet economy.

Sources: Chernobyl Disaster, at http://www.chernobyl.co.uk/health.html, consulted on 17 October 2005.

UNITED NATIONS
Office on Drugs and Crime

Consequences of the 1986 Chernobyl Nuclear Power Plant Accident (cont.)

➤ A plume of radioactive debris drifted over parts of the western Soviet Union, Eastern Europe, Scandinavia, UK and eastern USA. About 60% of the radioactive fallout came down in Belarus. The radioactive contamination of the Chernobyl accident was as much as 400 times the one of the Hiroshima bomb.

➤ The average thyroid cancer rate which was 4-6 incidents per million for Ukrainian children between 1981–1985 and rose to 45 incidents per million for the period 1986–1997

➤ Over 2,500 deaths were caused by the Chernobyl accident subsequently, according to the Ukraine Radiological Institute; a 2005 UN report attributed 56 deaths (47 accident workers and 9 children with thyroid cancer) to the accident, estimating that around 4,000 people will ultimately die from accident-related illness.

Sources: Chernobyl Disaster, at http://www.chernobyl.co.uk/health.html, consulted on 17 October 2005.

UNITED NATIONS
Office on Drugs and Crime

The 1987 Goiania (Brazil) Cesium-137 Accidental Leakage

➤ 90 grams of Cesium (1,375 Curies) were leaked from abandoned medical equipment

➤ 4 people died, 1 person had an arm amputated, 28 people suffered radiation burns and 54 were hospitalized for serious contamination

➤ 190 others were found contaminated

➤ Between 34,000 and 112,000 (of the town's 800,000 people) people were examined for contamination

Source: Alex Neifert. Case Study: Accidental Leakage of Cesium-137 in Goiania, Brazil, in 1987.
http://www.nbc-med.org/SiteContent/MedRef/OnlineRef/CaseStudies/csgoiania.html, consulted on 10 Oct. 2005; IAEA. Dosimetric and medical aspects of the radiological accident in Goiania in 1987. Vienna, IAEA, June 1998 ; Charles D. Ferguson, Tahseen Kazi and Judith Perera. Commercial Radioactive Sources: Surveying the Security Risks. Monterey, Center for Nonproliferation Studies, 2003, p. 22.

UNITED NATIONS
Office on Drugs and Crime

The 1987 Goiania (Brazil) Cesium-137 Accidental Leakage (cont.)

➢ 621 people were recognized as long-term victims by 2005

➢ A land area of about one square kilometre (roughly 40 city blocks) required a massive cleanup effort. Seven homes and some other buildings had to be demolished. 3,500 cubic meters of radioactive waste were generated. About 1200 of the 1375 curies could be recovered.

➢ Cleanup costs amounted to more than $ 20 million while the economic loss from collapse of the tourism industry and business were estimated to amount up to hundreds of millions of dollars.

Source: Alex Neifert. Case Study: Accidental Leakage of Cesium-137 in Goiania, Brazil, in 1987.
http://www.nbc-med.org/SiteContent/MedRef/OnlineRef/CaseStudies/csgoiania.html, consulted on 10 Oct. 2005; IAEA. Dosimetric and medical aspects of the radiological accident in Goiania in 1987. Vienna, IAEA, June 1998 ; Charles D. Ferguson, Tahseen Kazi and Judith Perera. Commercial Radioactive Sources: Surveying the Security Risks. Monterey, Center for Nonproliferation Studies, 2003, p. 22.

UNITED NATIONS
Office on Drugs and Crime

Shortcomings of TOPOFF 2 simulated Radioactive Attack, Seattle, May 2003

➢ Inability to share information due to shortage of secure phone lines and lack of security clearances by some participants

➢ Unclear government procedures

➢ Conflict across the chain of command

➢ Inability of government to provide consistent and accurate information about the path of radiological plumes (despite the fact that the exercise was heavily scripted

➢ Need for a radiological decontamination standard

Source: Charles D. Ferguson, William C. Potter et al. The Four Faces of Nuclear Terrorism. Monterey, Center for Nonproliferation Studies, 2004, pp. 296, 307.

UNITED NATIONS
Office on Drugs and Crime

Attempted & Aborted Uses of RDDs

➢ Late 1960s: A group of Nigerian scientists in Europe collected radioactive materials which they wanted to explode in Lagos, Nigeria to revenge the genocide on the Ibo. The material was "lost" in Portugal during shipment.

➢ 1970s: An employee of a nuclear facility attempted to slowly kill another man by placing highly radioactive metal rods under the seat of his car. The plot was aborted in time.

➢ 1970s: Dynamite was found at the Wisconsin-Michigan Power Company, a licensed nuclear facility in Point Beach (USA).

➢ 12 November 1972: 3 men with guns and grenades hijacked a Southern Airlines DC-9 and threatened to crash it into a reactor at the Oak Ridge National Laboratory in Tennessee if their $10 million ransom demand was not met. The hijackers circled over the installation and the personnel were evacuated before the hijackers' bluff was called.

Source: Database on Significant Nuclear and Radiological Incidents, Events, and Threats. Leiden, PIOOM, 1999 (updated last in May 2005); Rensselaer W. Lee, Smuggling Armageddon: The Nuclear Black Market in the Former Soviet Union and Europe, St. Martin's, 1998.

UNITED NATIONS
Office on Drugs and Crime

Attempted & Aborted Uses of RDDs (cont.)

➢ 17 April 1974: On a Rome-Vienna train a man, who wanted to protest against the treatment of the mentally ill in Austria (he himself had a history of insanity), dispersed radioactive material on passenger seats. 12 persons were contaminated but the level of radiation posed no danger to their health.

➢ 1993: A "Bosnia Front" threatened to detonate nuclear explosive devices in European cities lest certain political demands of Bosnia were met.

➢ 1993: A radioactive substance was planted in the chair of Vladimir Kaplun, director of a Russian packing company; over several weeks Kaplun was diagnosed with radiation sickness and died.

Source: Database on Significant Nuclear and Radiological Incidents, Events, and Threats. Leiden, PIOOM, 1999 (updated last in May 2005); Rensselaer W. Lee, Smuggling Armageddon: The Nuclear Black Market in the Former Soviet Union and Europe, St. Martin's, 1998.

UNITED NATIONS
Office on Drugs and Crime

Generic Terrorist Scenarios Involving Radioactive Materials.
In Increasing Order of Potential Consequences.

➤ *Low Consequences*

- Demonstrative symbolic minor acts of sabotage of nuclear facilities;

- (Threat of) Dispersal of small quantities of radioactive materials by terrorist or organized crime group, combined with demands and threats of larger dispersal if demands are not met.

UNITED NATIONS
Office on Drugs and Crime

Generic Terrorist Scenarios Involving Radioactive Materials.
In Increasing Order of Potential Consequences (cont.).

➤ *Medium Consequences*

- Silent dispersal of significant quantities of radioactive materials in subway or skyscraper ventilation system

- Attack on nuclear (rail-, highway, or sea-) convoy in transit resulting in release of radioactivity.

UNITED NATIONS
Office on Drugs and Crime

Generic Terrorist Scenarios Involving Radioactive Materials.
In Increasing Order of Potential Consequence (cont.).

➤ *High Consequences*

- Firebomb (high explosives and incendiary material) attached to a significant quantity of Cobalt-60 or Cesium-137;

- Siege-and-hostage occupation of nuclear reactor with or without the help of insiders and conditional threat of harm and destruction;

- Failed attempt to explode plutonium oxide IND resulting in non-fission dispersal of Pu-239;

- Suicide attack on nuclear power plant or storage sites for spent nuclear fuel suicide commando or by crashing a plane into a nuclear facility.

UNITED NATIONS
Office on Drugs and Crime

The Water Pollution Scenario

Low Consequences

Radioactive material could be placed by terrorist into specific enemy water supplies where they might be more effective than in large reservoirs. This was one of the uses considered by the U.S. military in World War II. However, some radioactive materials are not water-soluble; others, like Pu-238 would sink to the bottom of a reservoir. In any case, enormous quantities of an RDD would be needed to effectively contaminate public water reservoirs. In practice, this method works only against water storage tanks of individual buildings.

Source: Charles D. Ferguson et. Al. Commercial Radioactive Sources, op. cit., p.19.

UNITED NATIONS
Office on Drugs and Crime

The Silent Indoor Dispersal Scenario

Low Consequences

Dispersal of radioactive material for environmental contamination: radioactive material (Cs-137, Sr-90, Co-60) could be injected into the central ventilation system of major civilian structures (e.g. airport, medical centre, subway, office complex, shopping mall), potentially causing radiation exposure of inhabitants and users. This would require atomising the radioactive material where the particle size need would to be calibrated so that it could be inhaled by the prospective victims.

Source: F. Steinhäusler, M. Bremer-Maerli, L. Zaitseva. Assessment of the Threat from Diverted Radioactive Material and "Orphan Sources"- An International Comparison. In: IAEA. Measures to Prevent, Intercept and Respond to Illicit Material and Radioactive Sources. Vienna, IAEA, 2002 (C&SpapersSeries 12/P)., p. 275.

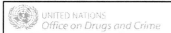

UNITED NATIONS
Office on Drugs and Crime

The Gas-Enhanced Blast Scenario

Medium Consequences

A terrorist group, having obtained 600 grammes of Plutonium (Pu 238) uses it for an explosive radiological dispersal. The two-stage explosive device attached to the Plutonium is meant first to puncture a 1,000 gallon or larger propane storage tank and, a moment later detonate the expanding propane cloud in the middle of a working day in a metropolitan area. A newspaper receives a call from the terrorist organization, indicating that radioactive material was dispersed with the explosion. Measurements indicate that 60 REM in an area 500 meters wide and 1000 meters long; 30 REM in a sectors 2 km wide and 1 km long; 15 REM in a sector 5 km long by 2 km and 10 REM in a sector 8 km long and 3 km wide.

Source: George Buck. Preparing for Terrorism. An Emergency Services Guide. Albany, N.Y.,Delmar Publishers, 1998,pp. 196/7. – REM is the special unit of dose equivalent which equals the absorbed dose multiplied by the quality factor (100 REM equals 1 Sievert).

UNITED NATIONS
Office on Drugs and Crime

The Manhattan Scenario

High Consequences:

A terrorist group with insider help acquires radiological materials, and detonates a dirty bomb in New York's financial district, dispersing radiation across a 60-square block area.

The Federation of American Scientists estimated that such an attack with Americium-241 could cost, in terms of decontamination and rebuilding, over $50 billion (less if rebuilding is not necessary).

Source: Testimony of Dr. Henry Kelly before the U.S. Senate Committee on Foreign Relations, 6 March 2002.. – Americium can be found in many smoke detectors but it would take millions of smoke detectors to construct a sizeable RDD. – In August 2004, British officials had arrested suspects who had plans to attack targets in London, including the Heathrow express rail line. Those arrested had a cache of household smoke detectors that the group wanted to cannibalize for the Americium-241 which they apparently thought was enough for a " dirty bomb". Reuters, 3 Oct. 2004.

UNITED NATIONS
Office on Drugs and Crime

The sabotage of a Nuclear Power Plant Scenario

High Consequences:

With or without insider help, several suicidal terrorist commandos familiar with its design and operation attack a nuclear power station, break through plant security systems and gain access to the reactor containment itself and exploit existing vulnerabilities to cause uncontrolled release of radioactivity into the environment.

RDD Attack Consequence Assessment Matrix

Types of Consequences	Level of Consequence		
	Low	Medium	High
Health	(Rems:0-100) Zero to temporary illness	(Rems:100-500) Immediate illness; up to 50% of victims will die	(Rems:600-500) Up to 80-100% fatality of affected population
Psycho-Social	Evacuation is efficient and orderly/low psychological damage	Loss of control of evacuation but regain control in a few days/ psychological impact is apparent	Permanent population shift and serious business loss
Political	Effective government response and social cohesion not affected	Response displays incompetence/loss of faith in government	Change in political system
Economic	Local, temporary and decontamination costs are moderate	National, prolonged and decontamination costs are substantial	International and decontamination costs are very high
Environmental	Local and of short duration	Local and of long duration	Regional and of long duration

Main points of Code of Conduct on the Safety and Security of Radioactive Sources

➢ Ensure that those who seek to possess radioactive sources are authorized to do so by competent regulatory authorities

➢ Require adequate safety and security of radioactive sources throughout their life-cycles from production to use to disposal

➢ Establish confidential, national registries of holders of sources

➢ Ensure that inventory controls are conducted periodically by licensees of the radioactive sources

➢ Conduct regular announced and unannounced inspections of licensees' facilities where radioactive sources reside

Source: cit. Charles D. Ferguson and William C. Potter. The Four Faces of Nuclear Terrorism. Monterey, Center for Nonproliferation Studies, 2004, p. 285.

UNITED NATIONS
Office on Drugs and Crime

Effects of Radiological Dispersion Devices

➢ RDD are not weapons of mass destruction

➢ Negative health effects of RDDs (beyond the harm caused by the conventional explosive blast) are minor in short run and modest in the long run

➢ Socio-psychological effects (panic, unorderly evacuation) can be substantial but can be mitigated by public information before and during incident

➢ Political effects might be significant

➢ Environmental effects are manageable

➢ Economic effects (disruptive area denial, decontamination and reconstruction costs) are likely to be the main effects and can be very substantial, amounting to billions of dollars

UNITED NATIONS
Office on Drugs and Crime

Terrorism Prevention Branch
Division for Treaty Affairs
United Nations Office for Drugs and Crime
P.O. Box 500
A – 1400 Vienna, Austria
Tel: + 43 1 26060 4278
Fax: + 43 1 26060 5968

3 RADIOLOGICAL AND NUCLEAR TERRORISM

David Martin
Director, Strategic Defense Programs, Tetra Tech EC Inc., USA

What is the threat?

- **Terrorists:**
  - Al Queda openly seeking WMD
  - Al Queda receiving Fatwa to use WMD against U.S. (Abu Shihab El-Kandahari, 26 December 2002)
  - Al Queda linked (financed by Al Queda) U.S. national Jose Padilla seeks to use RDD in Chicago
  - Have varying levels of smuggling experience, capabilities, and motivations

W.D. Wood and D.M. Robinson (eds.), *International Approaches to Securing Radioactive Sources Against Terrorism*,

Global Scale

- Terrorists traveling continents away just to get in the mix (scrum)
- Availability of radiological and to a much lesser extent nuclear material
- Knowledge:
  - Al Queda is systematic in study of targets
  - Constantly adjusting and refining tactics
  - Has sought and acquired WMD information (a London Times reporter discovers a blueprint for a "Nagasaki bomb" in files in abandoned al-Queda house in Kabul, Afghanistan, 19 November 2001)
- Chechen rebels who are believed to have ties to Al Queda have already demonstrated RDD capability

Transportation of Materials to Buyer

- Who is in possession of the material?
  - Less experienced smugglers - likely to go on their own.
  - More experienced - one or more couriers?
- What path will they choose?
  - Less experienced - common, legal routes
  - More experienced - established smuggling routes? (May still involve legal border crossings.)
- How will carrier respond if discovered?

IAEA Illicit Trafficking Database (ITDB)

- Most policy makers are aware of IAEA Illicit Trafficking Database

- They use it as a baseline to judge threats related to availability of nuclear and radiological material

- Tremendously useful as an overview to inform member states

- Illicit trafficking is still largely under-reported

- This has led to false sense of security based on the assumption of what we know because of the scope of the IAEA database

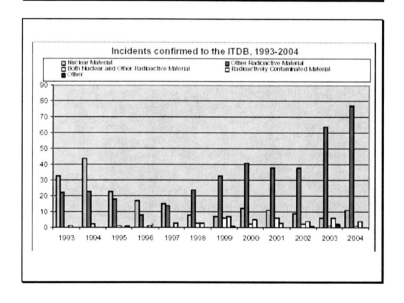

Incidents confirmed to the ITDB, 1993-2004

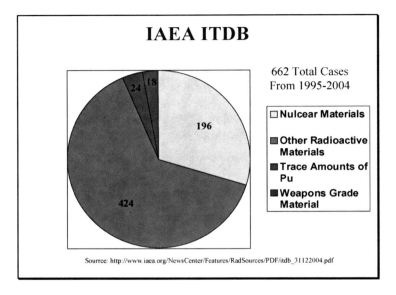

IAEA ITDB

662 Total Cases
From 1995-2004

- ☐ Nulcear Materials
- ◧ Other Radioactive Materials
- ■ Trace Amounts of Pu
- ■ Weapons Grade Material

Sourrce: http://www.iaea.org/NewsCenter/Features/RadSources/PDF/itdb_31122004.pdf

IAEA ITDB vs. Case data from Eastern Europe

Poland	No cases prior to 1994 when Poland began deploying radiation detection portal monitors	Since complete border coverage was achieved 12,000-16,000 cases/year
Russia	No cases prior to 1996 when Russia began deploying portal monitors	Since 2000 more than 1000 responses to alarms per year

It is a quantifiable military threat?

- Border enforcement is usually not prepared to deal with interdiction of potential IND or RDD (they have proved efficient at counter illicit trafficking in materials)

- Successful interdiction of nuclear or radiological smuggling cases are of **strategic** military interest

- Consequences of successful radiological or nuclear terrorist attack could equate to large natural disaster triggering calls for the military to provide support to civil authorities

- Depending on the location, multiple nations may be affected triggering coordination requirements between civil and military leaders

RDD Response may require Military/Alliance Involvement

- A radiological dispersal device, if well constructed and deployed, can affect a large metropolitan area:
  - By forced evacuations (population dislocation)
  - Lingering clean-up
  - Economic loss due to and during the clean-up
  - And a general breakdown in public order

- Depending on the scale of the event one nation's military may not be enough to support civil authorities and maintain operations thru recovery

Policy cannot be framed
from the viewpoint of 5-10 years ago

- No longer primarily opportunistic and financially motivated smugglers
- Securing nuclear facilities and more recently sources is by itself is not enough
- Not just a problem for Russia or Eastern Europe
- Criminal intelligence alone is not enough

Layered Defense/Defense in Depth

- Enhanced security measures for facilities, borders, and transportation networks
- Technical means and training for border enforcement officials
- Information sharing
- Military Counter-terrorism support to civil authorities in the event of interdictions of IND or RDD
- Response planning and exercises for catastrophic attacks including military and alliance support to civil authorities

Why Border Monitoring?

- Short of direct/tactical intelligence alerting a nation to the intended deployment and use of a nuclear/radiological device, border enforcement activity, including the use of radiation detection technology, may provide a country with its most effective "early warning system" along with a clear opportunity for interdiction prior to infiltration of the device or its components.

Technology as a Tool: Not a Panacea

Radiation detection equipment:

- Technical means for law enforcement to detect, identify, and interdict nuclear and radiological smuggling
- Technical means for military to identify and interdict suspect cargos
- Technical means for civil and military authorities to respond to radiological or nuclear attack

Equipment and Operations

- Radiation detection technology must be capable and associated with advanced training

- Confidence and integrity in performance of equipment is vital

- Traditional "Cold War" era military missions for response in event of nuclear exchange must be updated to include interdiction missions, coordination with civil authorities, and post-event response

Combat Operations: Not Enough by Themselves

- While so called tip-of-the-spear offensive operations against terrorism are critical to the success of the Global War on Terror

- Defensive preparations such as enhanced border security and intelligence operations are required to allow the detection and disruption of terrorist efforts already underway

What is being done already?

- UN Resolution 1540

- G-8

- IAEA Board of Governors launch Nuclear Security Initiative

- Unilateral and Bilateral efforts to enhance border security

- Enhanced transportation security (WCO, ISPS code, etc.)

- NATO WMD Center

- NATO Operation Active Endeavor (Maritime patrolling of the Mediterranean)

UN Resolution 1540

- The resolution obligates all states to develop and maintain "appropriate effective measures to account for and secure," nuclear, chemical, biological weapons, related materials, and their means of delivery, as well as "appropriate effective physical protection measures."

- The resolution also requires all states to put in place "appropriate effective" export controls, and to "develop and maintain appropriate effective border controls and law enforcement efforts to detect, deter, prevent and combat, including through international cooperation when necessary, the illicit trafficking and brokering," in such items.

Is it Enough?

- Most of these initiatives are necessary and contribute to the strengthening of international security

- Implementation remains slow and in some cases funding is a major constraint

- Translating policy into practical applications is a challenge

- Unilateral efforts to implement enhanced border security are spotty at best with the U.S., U.K, Russia, Poland, the Netherlands and others leading by example

- Military reform to reflect the changing mission environment has begun but is far from over

What should be Accelerated or Prioritized?

- Enhanced border security (set up early warning system through early detection)

- Coordination of border enforcement information with relevant military commands

- Military training for missions supporting civil authorities in interdiction of WMD & response to RDD/IND attack

  - New missions require new tool sets (i.e. WMD search equipment specifically adapted for maritime search, rapid deployment, and temporary checkpoints)

  - Stockpiles of decontamination equipment should be prepared to deal with mass decontamination support

  - Search tactics that do not burden commerce unnecessarily but support the mission (don't just deter - detect)

SECTION IV

COORDINATING RESPONSES TO RADIOACTIVE TERRORISM

1 KEEPING THE TERRORIST TRAGEDIES OF YESTERDAY FROM BECOMING THE TERRORIST CATASTROPHIES OF TOMORROW

Dr. John Hnatio

Executive Director, Institute for Complexity Management, Former Professor of Strategic Leadership and Decision-Making, National Defense University, USA

How the international community deals with future disasters, natural or man-made, is a critical topic of concern to the world.

Regardless of opinions one way or another, there is one thing that I am sure all of us agree on: we need to think about new and better ways to anticipate and prevent disasters and, if the worst still happens, how do we prevent disasters from becoming catastrophes?

Today, we must not only think about the possibility of terrorists using biological, chemical and radiological weapons of mass destruction to intentionally cause disasters that are designed to create catastrophes. There are also natural disasters like tsunamis, hurricanes, earthquakes and a potential avian flu pandemic that we must think about.

In the end analysis, the question of what we can do to prevent disasters, whether they are man-made or occur naturally, from escalating to become catastrophes is one of the most important questions of our time. But, much of the work and investments we have made in disaster preparedness can serve us well in anticipating, preventing and mitigating the consequences of both man-made and natural disasters to prevent them from escalating into catastrophes. This includes concerns about the malevolent use of radiological dispersal devices.

W.D. Wood and D.M. Robinson (eds.), *International Approaches to Securing Radioactive Sources Against Terrorism*,
© Springer Science+Business Media B.V. 2009

Background

But first a brief description about ourselves. The Institute for Complexity Management is a nonprofit organization dedicated to the mission of helping government and industry more effectively manage complex contingencies

The focus of our work is to conduct research and support educational activities involving disaster preparedness and risk management, response and mitigation.

To help us turn our research into practical applications we have formed relationships with a number of significant organizations including:

- MetroStar Systems
- Argonne National Laboratory (non-profit)
- Executive Development Associates
- Salus International
- Cyntelix
- Complexity Key (non-profit)

What We Have Heard Already …

- Prevalence: Radiological sources are everywhere. And it's not just a question of looking at international borders. We need to look at the problem in an integrated way – looking at policy, procedures, standards, the technology to deal with the problem and also the political will necessary to do things.

- "Wish for the best, but plan for the worst" as Margaret Thatcher once said.

- There is no epistemological base or structure for thinking about worst case scenarios. Everyone jumps to the worst case scenario –

but have they really thought about it? It's the difference between a disaster and a catastrophe. A catastrophe, in my opinion, is two or three simultaneous hits in places such as New York City and Wall Street, the port of Los Angeles and two or three other places in the United States, all done in a coordinated fashion. At that scale there's an effect on the whole economy of the country. There's a huge difference between taking care of one little dispersal device and the effect of a scenario where someone is committed to destroy you. That's one of the things we've seen with extremism – extremism doesn't necessarily have the same end as the old brand of terrorism used to have. Extremism means you really want to destroy your opponent in the sense of a war. So the distinction between radiological dispersal devices and weapons of mass destruction becomes very blurred. Because if they destroy my society, it doesn't matter what they use.

- We really don't know what we don't know.

- Intervention after the attack – very important. But there are also things that you can do before the attack, and we'll come to that later.

- The real name of the game is fear.

- Beware of the nuclear paparazzi because they are going to play on the fear. They're going play on the fear and drive our political masters into the worst-case scenario every time.

- Are we planning for the right things? We're doing a lot of things with technology – but are we putting them together in the right way?

- No international actions can ever replace a country's own actions – especially in the case of the United States where we have a huge number of these radiological sources.

The Threat Is Very Real

This shows the distribution, in May 2004, of Improvised Explosive Devices in the Baghdad corridor. The point is, these guys are using a whole bunch of different stuff – and they're getting better at it. They started with very simple devices – today their sophistication has taken quantum leaps.

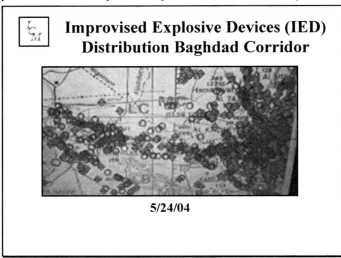

Improvised Explosive Devices (IED) Distribution Baghdad Corridor

5/24/04

What we're dealing with is a complex, adaptive situation. The terrorist is constantly looking at us, looking at what we do, and is constantly looking for the new weak underbelly.

The reason I mention this is there might be a lesson here for Radiological Dispersal Devices – what we see and expect from the terrorist today may be completely different a year from now. They're going to adapt, continuously change and do different things.

The bottom line is: if the bad guys really want it, they'll probably get it, and they will really use it.

We have already heard from a number of distinguished speakers and, I am sure, we will hear more during the remainder of this conference about the threat of terrorism and the very real potential for radiological terrorism.

The threat of terrorism has already been well-established. So I will try and deal with the practical question of how we can do a better job of detecting, deterring, preventing and mitigating the consequences of terrorist attacks including the malevolent use of radiological dispersal devices.

The systems we have developed at the Institute for Complexity Management are based on research conducted at National Defense University and at The George Washington University. The research is based on achieving a better understanding of how complex adaptive systems work and using this knowledge to help us more effectively anticipate, prevent and mitigate the consequences of terrorist attacks.

**What we have learned
based on experience...**

- Twin Towers I (1993)
- Oklahoma City (1995)
- U.S. Embassies (1998)
- Cerro Grande Fire (2000)
- "Big City" Simulation (2000)
- Twin Towers II (2001)
- Hurricane Katrina (2005)

Being prepared for the worst is a very hard thing to do

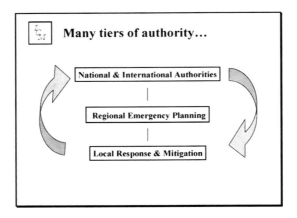

The process is continuous and is designed to cut across the entire system both vertically and horizontally. Going through different simulations to develop better response and mitigation plans and more informed decision makers who agree on who is responsible for what before a disaster strikes is the key to preventing disasters from becoming catastrophes. Good disaster preparedness planning is a continuing process and requires the investment of people's time and resources. But it, too, is key to preventing disasters form becoming catastrophes.

Moving in the Direction of an International Counter-terrorist Strategy Will Require …

Moving from this …

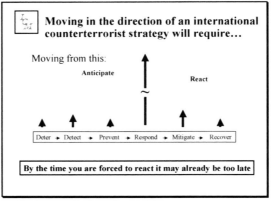

To this …

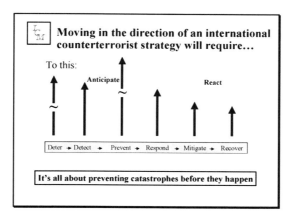

But from the lessons we have learned from events such as 9/11, Hurricane Katrina and other man-made and natural disasters, our current center of gravity is one that focuses on reaction after the fact. At the Institute for Complexity Management our research leads us to believe that we must shift to a new center of gravity, one that is based on anticipation and preparation. We believe that a new center of gravity that focuses on much better preparation as the priority has got to be the new focal point of any effective counterterrorist or disaster preparedness strategy.

At ICM our studies lead us to conclude that any effective disaster preparedness strategy designed to prevent disasters from becoming catastrophes must be based on the six critical factors of:

- Thinking about the unthinkable
- Thinking about the unthinkable before it happens as opposed to during or after the catastrophe has occurred
- Systematically thinking through the critical decisions, actions and responsibilities of the different players before events happen
- Achieving consensus on who is going to be responsible for what
- Systematically preparing to respond and mitigate the consequences of a disaster
- Long term economic recovery

The bottom line: it's all about preventing disasters from happening in the first place and, if they do happen, keeping them from becoming catastrophes.

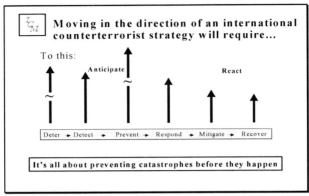

At ICM, we believe that there must be a shift in the center of gravity from 'respond and react' to 'anticipate, prepare and mitigate'. In the end analysis, effective response depends on the investments you make in preparation. The more you invest in mitigation, the shorter and more cost effective long term recovery becomes.

So, What We Can Do to Prevent Terrorist Tragedies from Becoming Catastrophes?

- Think about events before they happen, not during and afterwards to anticipate the unexpected.

- Make sure that a consensus exists among key decision makers at all levels on exactly who will do what, when and how it will be done before disasters strike.

- Think about events beforehand in a structured way to develop plans and achieve consensus among decision makers.

- Separate the quantitative reality from qualitative political and social process.

- Think about all of the possibilities from multidisciplinary frames of reference.

- Continuously ask yourself, "What is it that had I known before the catastrophe occurred could I have used to prevent it from happening in the first place?
- Make certain that there is effective communication among leaders at all levels
  - Bring the right information, at the right time, in the right form so that decision makers can use it effectively.
- Recognize human limitations in dealing with complex events
  - Conduct science-based simulations that use specialized disaster Decision Support Systems (DSS).
  - Constantly update disaster DSS.
  - Use technology in a way that it helps our decision makers.

How the CSM Method Works?

At ICM we've been working with our partners to create a practical process that uses Decision Support Systems to deal with a range of complex contingencies the nation is likely to face in the coming years.

Here's a brief description of the five major elements of the process:

1. Use simulations to help us think about "it" before the real thing happens; build simulations with everyone's inputs; make sure they are scientifically accurate.
2. Bring multidisciplinary decision makers and technical experts together
   - Cut both vertically and horizontally across the system to develop simulations.
   - Bring decision makers and experts from all levels of government together to run through simulated events.
3. Capture their ideas; determine critical decisions and best decision options; achieve consensus around best decision options before the event occurs; consider both the expected and the unexpected
   - Don't let qualitative political process stand in the way of quantitative reality.

4. Build a knowledge library with these inputs that gets smarter and smarter as more simulations are conducted.

5. Use the knowledge library to help our decision makers manage disasters more effectively so they don't become catastrophes.

Now, some of you may be thinking, "Heah, we already do table top exercises, why don't we just do more of them and we'll be fine."

CSM Decision Support Systems

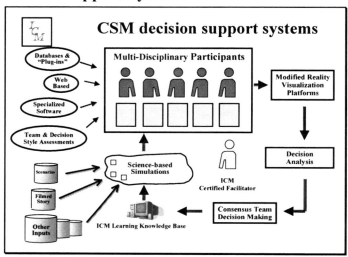

The process begins with the development of simulations. We work the entire system both vertically and horizontally, picking the brains of experts at every level. We identify pre-cursor problems that can lead to disasters or cause disasters to escalate to become catastrophes. We identify critical decision points in responding to simulated disasters and look at the range of possible decisions that could be made and their extended order effects. We build models that help us show the extended order effects of decisions. We also ask experts a critical question as they build science based scenarios of terrorist attacks, namely, what is that I had known before the attack could I have used to interdict the "bad guys" and have prevented the incident from happening in the first place. These become what we call the "indicators" and "warnings" of terrorist attacks and are structured or indexed as part of a supporting knowledgebase. These indicators and warnings can be used to develop focused intelligence collection strategies and to data harvest information from "the inside out."

Based on this extensive preliminary work, we bring a select combination of decision makers, first responders and multidisciplinary subject matter experts together to run through our simulations. We use a number of tools and techniques to help the group "reverse engineer" critical decisions and decide on best decision options under a given set of circumstances. We use a special process to help them achieve consensus on the best decision to make. We structure different options using decision trees based on these inputs that can be archived in a supporting computer knowledgebase and that gets "smarter and smarter" as successive groups run through the simulation.

The resulting knowledge base can be used for educational, strategic and tactical operational uses as a planning and response tool to help manage similar events that confront decision makers in the real world.

Modified Reality Visualization Platforms

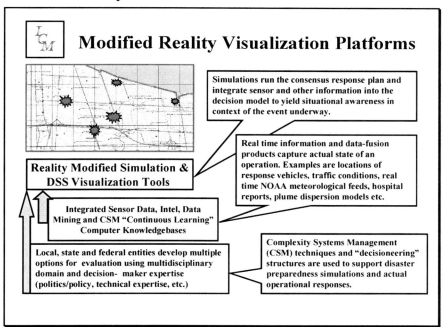

We use special technology tools to help decision makers "play through" hypothetical disasters. We use reality modified visualization and decision support system visualization tools during simulations to produce consensus response plans; we use real time data and data fusion products, e.g., location of victims, population density, traffic patterns, location of response and evacuation vehicles, hospitals, NOAA real time meteorological data, plume dispersion models, etc. Science based models and other techniques are used to help our decision makers achieve consensus on "best" decisions and to capture their rationale for selecting one decision option in lieu of another. The results are structured and archived in a supporting knowledge base that gets "smarter" as subsequent simulations are run. Knowledge-bases are designed so that they can also be used to support decision makers during actual operational responses to disasters.

We use critical infrastructure mapping and geospatial imagery of critical infrastructures …

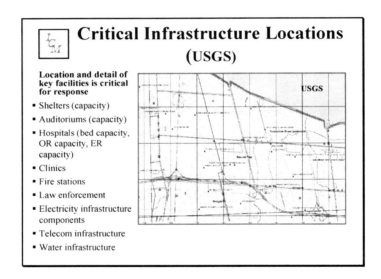

Models ranging from the most advanced weather modeling, flood modeling, plume modeling with real time NOAA meteorological feeds, and other models constructed for specific applications are used to support modified reality visualization platforms and ICM decision support systems.

Weather Modeling Analysis (NASA)

Region-specific detailed weather modeling

- Precipitation
- Wind
- Duration

Hydrology and topography modeling …

 Hydrology/Topography Modeling (ESRI)

- Detailed terrain analysis for hydrology and topographical analysis of likely flooding

- Coupled with weather modeling for severity of flooding

- Subject-matter expertise from Army Corps of Engineers

Population density modeling …

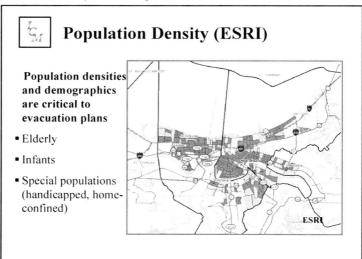

Geospatial imagery for damage assessment …

Next Generation Disaster Preparedness Decision Support Systems

In the future, our leaders will need better ways to anticipate, prepare for and reach consensus on protecting the complex interweaving infrastructures that bind our world together. They will need to think about complex events before they happen and achieve consensus on what best to do, how best to do it, and who is best at doing what must be done. Technology will never replace the human intellect, but it can help us bring the right information, at the right time and in the right context to help decision makers prevent disasters from becoming catastrophes.

The failure to detect the Asian Tsunami, global warming and changing weather patterns, the failed response to Hurricane Katrina may just be some of the most recent examples of disasters gone bad.

Today, we must expect harsher more unpredictable weather patterns, the potential for a global avian flu pandemic, and manmade accidents, not to mention the burgeoning threat of nuclear, biological and chemical attacks by terrorists using weapons of mass destruction. As with the threat of radiological dispersal devices, the threat is international in scope.

We must bring the right people, organizations and technology together in new ways if we ever hope to prevent future disasters from becoming catastrophes that can threaten global stability.

Why Not Just Do More Table Top Exercises?

Well, at both National Defense University and The George Washington University we have found that there are some serious limitations to table top exercises.

Limitations such as:

- Scope of the exercise is based on time and availability of key decision makers and fear of attribution.
- Dependent on the knowledge of the controller/scenario writer.
- Past performance is usually used as the indicator of future performance.
- Too often quantitative reality gives way to entertainment.

- Cognitive dissonance drives people to "look for the low hanging fruit".
- Single "Newtonian" outcomes versus a range of extended order effects.
- No systematic indication of the physical "indicators" and warnings of impending attacks.
- Not devised to deal critically with the quality of decisions
    - Critical decision points are frequently not identified and "reverse engineered" to achieve "best" decisions based on multidisciplinary consensus.
    - Table tops mix up quantitative reality with qualitative political process and seldom result in realistic outcomes.
- Do not make full use of current technology capabilities
    - Capturing and structuring knowledge for future use
    - Metaphorical entertainment versus analogous science
    - Modified Reality visualization platforms

CSM Decision Support Systems

At ICM we have developed a process that brings the right people, organizations and technology together to address the weaknesses inherent in today's generation of table top exercises.

The process looks like this:

- Realistic numbers of decision makers and experts; non-attribution policy.

- Simulations are built on broad based social, scientific and technical multidisciplinary inputs
    - Build on past performance but designed to anticipate the unexpected.
    - Science based simulations and models.
- Human learning assessments to help people "to climb trees for the best fruit".
- Arrays of possible outcomes with ranges of extended order effects.
- Identify critical decision points
    - "Reverse engineer" critical decisions to achieve "best" decisions based on multidisciplinary consensus.
    - Separate quantitative reality from qualitative political process and deal with both.
- Capture and structure knowledge for future use via learning knowledge bases.
- Make full use of current technology capabilities
    - Science-based simulation
    - Modified reality visualization platforms
    - Decision mapping
    - Consensus Team Model

ICM Decision Analysis Technique

In ICM disaster preparedness simulations, we carefully and methodically identify those points in a simulated disaster where decision must be made – Max Mayfield's unambiguous prediction that Katrina would slam into the Gulf Coast near New Orleans a full two and a half days before the hurricane made landfall, is an example of what we would call a Critical Decision Point or CDP. When Mayfield tried to warn us there were several decisions we could have made. In ICM Decision Support environments, we "reverse engineer" CDP's to look at a range of possible alternatives for decision makers to consider as they choose and achieve consensus on "best" decision options. These decisions are used to create special event decision maps that can then be archived in a supporting computer knowledgebase with other critical information.

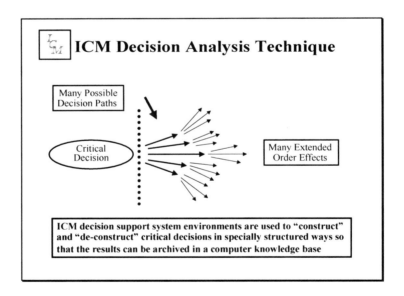

Getting people to agree on things can be very difficult if they come from different organizations with different responsibilities and perspectives and have different power bases. We have developed a special way to bring diverse decision makers together to reach consensus on "best" critical decisions in a systematic way. It is known as the Consensus Team Decision Process.

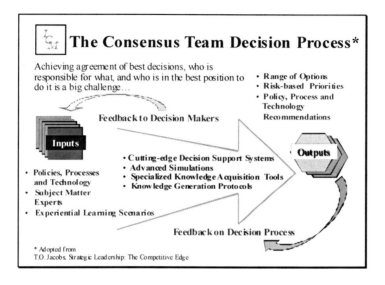

The Consensus Team Decision Process*

Achieving agreement of best decisions, who is responsible for what, and who is in the best position to do it is a big challenge...

- Range of Options
- Risk-based Priorities
- Policy, Process and Technology Recommendations

Feedback to Decision Makers

Inputs

- Policies, Processes and Technology
- Subject Matter Experts
- Experiential Learning Scenarios

- Cutting-edge Decision Support Systems
- Advanced Simulations
- Specialized Knowledge Acquisition Tools
- Knowledge Generation Protocols

Outputs

Feedback on Decision Process

* Adopted from
T.O. Jacobs, Strategic Leadership: The Competitive Edge

SECTION V
FUTURE SECURITY OF RADIOACTIVE SOURCES

1 LESSONS LEARNED, WEAK POINTS AND FUTURE IMPROVEMENTS

Ivan Gorinov
Head, Division of Nuclear Material and Physical Protection, Nuclear Regulatory Agency, Bulgaria

Security of Radioactive Sources – lessons learned, weak points and future improvements

Ivan Gorinov

Head, Division of Nuclear Material and Physical protection
Nuclear Regulatory Agency, Sofia, Bulgaria

W.D. Wood and D.M. Robinson (eds.), *International Approaches to Securing Radioactive Sources Against Terrorism*,
© Springer Science+Business Media B.V. 2009

This presentation provides information about:

- International co-operation in nuclear security - Bulgarian experience and lessons learned
- Weak points in international system for combating illicit trafficking of radioactive substances
- What should be changed in order to improve the security of radioactive sources

"Ideal" world

"Ideal" world

"Real" world

"Real" world

International co-operation

- Joint Bulgarian – Turkish border exercise
- International Radiological Threat Reduction (IRTR) Program – US DOE/NNSA
- IAEA/EU Project „Strengthening of State's Capabilities for Detection and Response to Illicit Trafficking

Joint Bulgarian–Turkish border exercise

IRTR Program – US DOE/NNSA

- Security upgrades of gamma irradiators and medical treatment facilities
- Criteria established by the NNSA - > 100 Ci for gamma and > 1 Ci for alpha sources – Transportation of problematic ownership orphan sources to the secure RAW facility

IAEA/EU Project
"Strengthening of State's Capabilities for Detection and Response to Illicit Trafficking"

- First project considering lessons learned from previous projects – need of uniformity and general implementation, not piece by piece
- Border monitoring equipment upgraded at Varna and Bourgas ports: 6 railway, 4 road and 8 pedestrian units. Additional request for airports at Sofia, Plovdiv, Varna and Bourgas, the border railway stations at Svilengrad, Kardam, Rousse, Kalotina and Kulata
- Training provided for law enforcement staff
- Equipment delivery in the end of October. Training started first week of November in Athens, Greece

Major concerns
Stolen but not recovered!

- Kremikovtzi case - 2002

  - ◆ two 111 GBq Cs-137 sources stolen
  - ◆ very dangerous (CsCl)
  - ◆ intentional act – who, why?
  - ◆ still not recovered
  - ◆ **serial numbers: 5173GN and 5174GN**

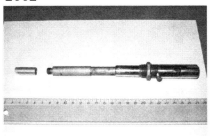

Major concerns - Orphan and "semi-orphan" sources

Empty depleted uranium containers had been found near the roadside by passing pedestrian – 2003
What if they were not empty ???

Major concerns - Orphan and "semi-orphan" sources

- Bankrupted or going to bankrupt enterprises, scientific institutes
  - ◆no money to secure and/or remove to safe place
  - ◆often try to simulate theft or intrusion
- Many type of sources are unnecessary now because of technology change – Pu and Am static electricity neutralizers and smoke detectors

Major concerns
Border control failures

- Railway checkpoints ???
- Coverage of ALL road checkpoints ???
- Example of "Zircon" case - **The border control systems of several European countries failed!**

Bulgarian border checkpoints

Major concerns
Co-ordination between authorities

- National Action Plan in case of illicit trafficking of NM & RM
  - responsibilities of different authorities
  - "who is the boss" - clear rules in order to determine who is the chief field executive officer
  - specific safety procedures
  - specific legal procedures - criminal investigation, witnesses, evidences, arrest, etc.
  - nuclear forensic analyses
  - criteria to trigger emergency plan
  - handling the media
- But is it working ? Example – "Zircon" case

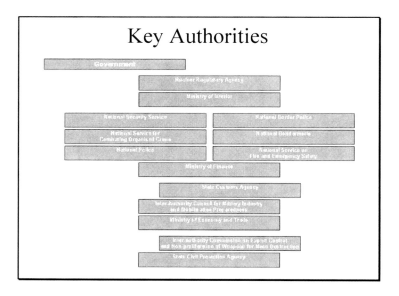

Legal concerns

- The illegal possession of radioactive source is still not recognized as crime in many countries – in Bulgaria it was included in the Criminal Code two years ago

What should be done?

- Orphan and "semi-orphan" sources
- Strengthening of border control
- Establishing of **REAL** co-ordination between state authorities
- Establishing of **REAL** co-ordination between states
- Implementation of DBT approach in the same way as for physical protection of nuclear facilities and nuclear material
- Security culture

Design basis Threat (DBT)

- What is DBT?

Definition in IAEA INFCIRC/225/Rev.4 (Corrected): *"The attributes and characteristics of potential insider and/or external adversaries who might attempt unauthorized removal of nuclear material or sabotage, against which a physical protection system is designed and evaluated."*

Security Culture

- What is "security culture"?
- How to convince people?
- Scientific world = world of freedom
- Target:
  - ◆ security measures should be adopted as "usual" and "necessary" measures by all people involved, for example – airport security
  - ◆ the politicians and high level management staff should understand that nuclear and radiological security is something really necessary. The politicians approach –"it is cool, and interesting for the public" and management approach – "a big burden" should be avoided.

What NATO could do?

- As a military alliance, NATO has a broad experience in joint commanding of military troops from different countries.
- Good standardization developed for planning and execution of operations.
- Such experience could be used for improving co-ordination between different countries.
- As NATO can bring together US, German, Bulgarian and other allies in Afghanistan it can provide the working model for international co-ordination in combating illicit trafficking
- To provide source information from NATO intelligence services to be used in DBT process

2 PROTECTING MAJOR PUBLIC EVENTS AGAINST NUCLEAR RADIOLOGICAL TERRORISM

Dr. Klaus Duftschmid
Professor for Radiation Protection, Technical University Graz. Consultant to Department of Safeguards and Technical Cooperation, International Atomic Energy Agency (IAEA), Austria

Based on material by: Rolf Arlt, IAEA

W.D. Wood and D.M. Robinson (eds.), *International Approaches to Securing Radioactive Sources Against Terrorism,*
© Springer Science+Business Media B.V. 2009

The Threat …

Bali, October 2002

Bali October 2002

The Threat …

Madrid, 2003

Several bombs   **> 200 victims**

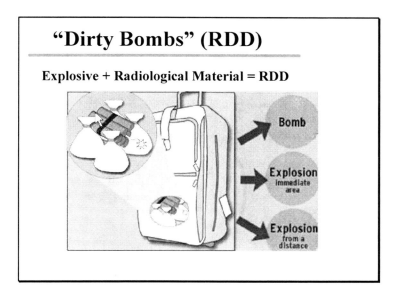

RDDs: Do they exist ?

History of Dirty Bombs

- 1995: Chechen rebels placed an RDD in the Izmailovsky Park in Moscow near the Kremlin (containing ^{137}Cs), it was disarmed
- 1998: a container of radioactive material was found attached to an explosive mine near a railway line not far from Angun in Chechnya; it was defused

Jose Padilla (Abdullah Al Mujahir) had allegedly been trained by Al-Qaida in Pakistan to build a radiological bomb and sent to the U.S. to locate possible targets.

Arrested in Chicago in May 2002. "They didn't think they'd have a problem getting radiological materials, but they didn't have any" a US official told the media.

Possible Consequences of RDDs

- **Case 1: Pu-smuggling from Reprocessing Plant Karlsruhe – Germany, July 2001**
  - **2 flasks contaminated with few mg Pu stolen**
  - **3 persons, 2 apartments and 3 cars contaminated**
  - **Decontamination took 1 year, costs > €1.5 m EUR**

- **Case 2: Goiania, Brasil, 1987: Theft of ^{137}Cs Teletherapy unit with 1375 Ci from scrap yard**
  - **4 people killed, 28 seriously burned, hundred contaminated, 112000 persons monitored**
  - **Panic**
  - **Total costs > $20 m USD**

The Threat ...

- **IAEA Director General, Dr. El Baradei on the Global Threat by Nuclear Terrorism:**
  - Terrorists are unconcerned about exposing themselves to radiation and could easily conceal a source in a truck, suitcase or on their body
  - The danger of handling powerful radioactive sources is no longer an effective deterrent, which dramatically changes previous assumptions !
  - "Security of nuclear and radiological material has taken on dramatically heightened significance in recent years in the work of IAEA"

Goals of Terrorists

- **A global dramatic response in the media is more important for their aims than killing a lot of people**
- **The psychological and economic effects of detonating a RDD in a large city are greater than multiple deaths**
- **Major public events, where large groups of people are gathered with strong participation of the media, enlarge the terroristic effect**

Assumptions for nuclear terrorism

- Assume following terrorist capabilities
  - Attack nuclear facilities and transports
  - Construct an effective RDD (dirty bomb)
  - Construct a crude nuclear explosive device (if terrorists acquire nuclear material assume that a simple gun-type device is within their capability)
- Abilities of present terrorist groups
  - Planning, global networks, co-ordination of attacks, decentralized operation (Osama owned a large globally active construction company)

- Resources
  - Financial (drugs and money laundering), information (internet), total "religious" commitment

How easy is it ?

Probability

- Theft of a nuclear weapon
- Acquisition of nuclear material to build a nuclear explosive device
- Attack on or sabotage of a nuclear facility or transport
- Malicious use of radioactive sources (RDD)

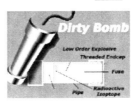

Nuclear Security of Public Events

Example: Olympic Games, Athens 2004

- Trilateral Project (GAEC, IAEA, USDOE)
- Considerable Budget: ~ 25 Mio EUR (most from US)

- Multiple Lines of Defense:

    1. Threat Assessment - Goal:
        - For nuclear or radiological material in the country: prevent attack on nuclear or other facility (sterilization plant, hospitals, industrial sources, transports, sources in scrap, waste storage for radioactive material)
        - Prevent nuclear or radiological materials illegally brought in the country or into the venues

Nuclear Security of Public Events ...

2. Physical Protection Measures
    · Research Reactor GRR1 "Demokritos"
    · Sterilization facility (^{60}Co source array):
    · 22 medical clinics in 16 hospitals

3. Border Monitoring
    · Installation of portal monitors: 6 airports, 11 Seaports, 10 land border crossings
    · Use of hand-held instruments: 181 γ PRDs, 32 RIDs, 30 $\alpha\beta\gamma$- survey meters, 10 NaI survey meters, 30 digital dosimeters

4. Training
    · 100+ courses, 1000+ trainees

Principles for Radiation Monitoring

- Locally deployed Radioisotope Identifiers (RIDs) to enable instant categorization of a radiation alarm
- Local clusters of mostly three security officers (Police, Army)
- Integration of highly sensitive radiation detectors (PRDs) into existing Security systems (metal detectors, X-ray units)
- Mobile expert teams on alert

Placing PRDs on X-ray units

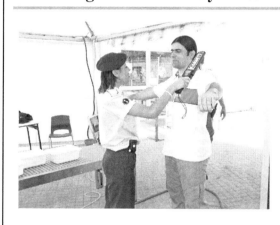

1: PRD 1
   on exit tray

2: PRD 2 worn
   on belt

Personal Radiation Detectors (PRD)

Low-end Gamma only units
Price ~ 1000 USD

Thermo RadEye   Polimaster PM1703M

Personal Radiation Detectors (PRD) ...

High-end Gamma/Neutron units
Price: ~ 2500 USD

MGP PRD-100   EXPLORANIUM GR-100

Radioisotope Identifiers (RID)

Presently most advanced RIDs
Price: ~ 15000 USD

Target Fieldspec   Exploranium GR-135

Radiation Portal Monitors (RPM)

Truck-, Car-, Pedestrian Monitors
Price range: ~ 25 – 150 k USD

RADOS RTM919N   EXPLORANIUM AT900

A New Category of Equipment

High sensitivity hand-held/mobile/fixed installed gamma/neutron radiation monitor

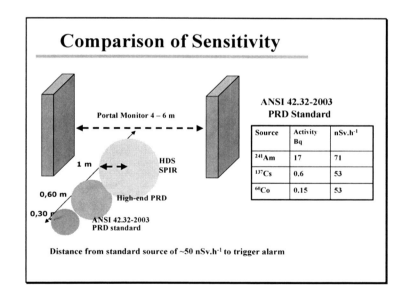

- a "cheap RPM"
- detection & search
- high sensitive CsI γ
- high sensitive ³He n
- networking
- GSM localization
- data storage

MGP HDS-100

MGP SPIR

Comparison of Sensitivity

Portal Monitor 4 – 6 m

1 m

HDS
SPIR

0,60 m

High-end PRD

0,30 m

ANSI 42.32-2003
PRD standard

Distance from standard source of ~50 nSv.h⁻¹ to trigger alarm

ANSI 42.32-2003 PRD Standard

Source	Activity Bq	nSv.h⁻¹
²⁴¹Am	17	71
¹³⁷Cs	0.6	53
⁶⁰Co	0.15	53

Problem: "Innocent" Alarms

- About 1 in 1000 persons receives medical radioisotopes for diagnostic and therapeutic purposes and may cause an "innocent" alarm
- At the Greek Olympics > 10 times a day
- After detection identification by RID is needed which causes delays and logistic problems
- Better: detection and identification at once
- Solution: Spectrometric Monitors

New spectrometric Monitor

Detection + isotope identification + innocent alarm rejection

- high γ detection sensitivity of large NaI detector
- instant identification and categorization
- Auto NORM and MED rejection
- masked and shielded SNM and RDD detection
- unattended network operation

MGP SPIR-IDENT

Spectral Gamma Scanner (IAEA)

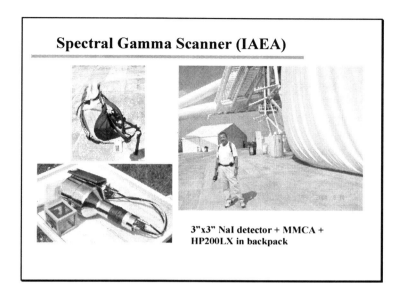

3"x3" NaI detector + MMCA +
HP200LX in backpack

The Future: Central Networks

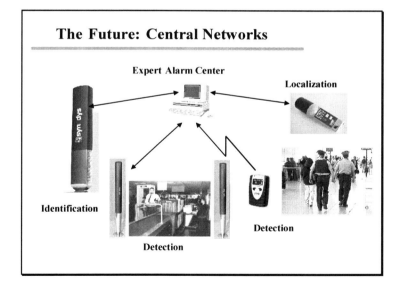

Expert Alarm Center

Localization

Identification

Detection

Detection

Conclusion

- The risk of actual deployment of nuclear and radiological weapons has considerably increased in the last years.
- There are no indications in the moment that terrorist groups have access to NBC weapons, however, Al-Qaida is definitely trying to get access.
- The required precautions are not primarily dependent on the probability of occurrence but on the risk, i.e. the product of probability times damage and the precautionary measures must be commeasurable to the risk.
- Radiological weapons ("dirty bombs") could easily be used by terrorist groups any time, suicide bombers don't worry about their own radiation exposure if they use strong sources.
- Technical measures are needed to detect illicit use of nuclear and radiological material and to improve homeland security in various situations, on borders, strategic points inside States and at many occasions such as public events, where large groups of people are gathered, with strong participation of the media, enlarging the terroristic effect.

3 SECURING RADIOACTIVE SOURCES AGAINST TERRORISM IN GEORGIA

Levan Gogua
Deputy Director, Service for Nuclear and Radioactive Safety, Georgia

W.D. Wood and D.M. Robinson (eds.), *International Approaches to Securing*
Radioactive Sources Against Terrorism,
© Springer Science+Business Media B.V. 2009

Georgia Has Extensive Legislation on Nuclear and Radiation Safety…

- Frame law "On Nuclear and Radiation Safety"
- Law "On Transport of Radioactive Substances"
- Law "On Radioactive Waste and Radioactive Waste Storage"
- Law "On Health"
- Decree "On register of Radioactive Waste"
- Decree "On Inspection of Nuclear and Radiation Activity"
- Decree "Limits for Radiation Protection"

… and in Addition, it Subscribes to International Agreements and Conventions

- Safeguard agreement between Georgia and IAEA for nuclear nonproliferation purposes.
- Additional Safeguard Protocol for nonproliferation purposes between Georgia and IAEA is also signed and adopted.

Organizations Responsible for Radiation Protection in Georgia

The following institutes are responsible for radiation safety according to Georgian frame law: "On Nuclear and Radiation Safety":

- Ministry of Environmental Protection and natural Recourses
- Ministry of Labor, Health and Social Defense
- Ministry of Internal Affairs
- State Technical Supervision Department
- National Center for Standardization and Certification
- Ministry of Defense
- Ministry of Economical Development

The Problems of Radiation Safety in Georgia

Georgia does not possess nuclear weapons or Nuclear Power Plants. Our one nuclear research reactor was stopped in 1987 because of public opinion. All fuel (fresh/used) was sent out of Georgia. The base of the reactor tank was covered by special concrete. And the reactor core and other high activity parts of the reactor were concreted over. Therefore Georgia is not a potential source for a nuclear disaster or nuclear threats to neighboring countries.

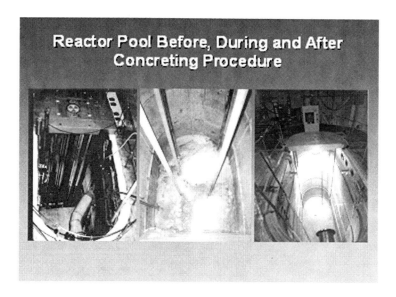

Reactor Pool Before, During and After Concreting Procedure

Meanwhile a lot of radiological accidents took place in Georgia. During the period 1995–2002, 17 people received serious injuries from "orphan" radioactive sources. There were five lethal events among them.

"Orphan" Radioactive Sources in Georgia

After the end of the Soviet Union a lot of radioactive sources were not under proper control. The main reasons for this situation were the following:

- Removal of Soviet military troops from Georgia.
- The procedure of enforced privatization at the end of 20th century in Georgia. Many enterprises owning radioactive sources changed their profile of activity. As a result many radioactive sources lost proper control.

Providing an appropriate level of physical protection and radiation safety for safekeeping of radioactive sources needs sufficient financial support. Therefore there were cases of sources being left without control, burning into soil, removal for scrap, the purchasing of containers together with sources, illicit trafficking and others.

Carrying Out Search and Recovery Operations for "Orphan" Radioactive Sources

- Removal of all found "orphan" radioactive sources from former Soviet military bases
- Decontamination of radioactively polluted areas of military bases, enterprises and medical organizations
- Gamma-Aero-testing of Georgian settled territories
- Gamma-testing by cars and pedestrians of other high mountain regions of Georgia

- Reconstruction and update of the storage of radioactive waste
- Establishing constant relation with population for gaining more complete information for searching of "orphan" radioactive sources
- Regular control of organizations handling radioactive sources
- Equipping of border checking and custom points with up-to-date detecting apparatus
- Participation of Georgian specialists for radiation protection in training courses, seminars, workshops organized by IAEA and other international organizations

There were found and recovered 263 "orphan" radioactive sources with activities from 1 mCi to 35,000 Ci since 1996. Seventy percent of them were found on the territories of military bases. More than half the number of found sources were naked, without any shielding. Twenty sources were found burned into the soil.

Illicit Trafficking of Nuclear Materials and Radioactive Sources

The large number of "orphan" radioactive sources and the geographical location of Georgia became temptation for smugglers to conduct illicit trafficking of nuclear materials and radioactive sources through Georgian territory. During the period 1999–2004 security guards caught smugglers trying to transfer to Turkey containers with Cs^{137} and Sr^{90}. The most important cases involving the seizure of nuclear materials at Georgian border included:

- U^{235} weight 998.87 g, enrichment 3.3% (1999)
- U^{235} weight 1,700 g, enrichment 3,4%
- U^{235} weight 918 g, enrichment 16%

The Possibility of Using of Nuclear Materials and Radioactive Sources for Terrorist Purposes

Check points at the border of Georgia are not equipped with appropriate detecting systems for nuclear and radioactive materials. There is a lack of specialists on the borders (custom and border guards) having high enough qualification for handling up-to-date detecting systems. Therefore there is some threat for the conduct of illicit trafficking of nuclear materials and radioactive sources through Georgian borders. The possibility of using radioactive sources for creation of so called "dirty" bombs is also to be considered.

The solution to all the problems related to illicit trafficking and the possible use of nuclear and radioactive materials for terrorist purposes is possible only in close collaboration with other countries. Therefore Georgia needs the assistance of developed countries in this task.

SECTION VI

UNCORKING THE BOTTLENECKS

1 BREAKING DOWN BARRIERS TO COOPERATION BETWEEN GOVERNMENTS, CIVIL SOCIETY AND BUSINESS

Michael McKay
CEO, McKay's International Corporate Communications & Public Affairs Management, Switzerland

New Perspectives on the Changing Role of Corporate Security

Although an 'outsider' to your profession, I have worked closely over many years with colleagues who handle the same difficult issues that you do, in a very tough industry: tobacco.

I recognise the character and blend of skills necessary for you to do your job successfully: intelligence, forcefulness, tact, resilience, tenacity, diplomacy, sagacity, and clairvoyance, to name just a few!

You are expected to have the gift of the futurologist when confronted by life-threatening situations caused by criminal or terrorist actions; while many of your colleagues are free to parade the doubtful wisdom of what our American friends call the 'Monday morning quarterback'. You do not have that luxury.

What can I contribute? What new perspectives can I offer you? You are the experts. You deal with the reality … daily.

Perhaps I am taking a huge risk by simply standing here. The only obstacle between you, cocktails and Basel by night!

But I'm here to take that risk, and hope you find what I have to say of interest. Interesting enough to stimulate some discussion before we take the antique tram.

W.D. Wood and D.M. Robinson (eds.), *International Approaches to Securing Radioactive Sources Against Terrorism*,
© Springer Science+Business Media B.V. 2009

I want to focus on three things:

1. Change and its implications for security
2. Threats: organised crime and terrorism
3. Corruption inside the firm

Then I'll finish by drawing some conclusions about the future, sharing some ideas with you that I hope will stimulate further thought and reflection.

1. Change and its Implications for Security

First, change: you don't need me to tell you how much the world has changed. But when was the watershed?

Was it 1989 and the collapse of the Berlin Wall?

Was it 9/11 in 2001: an audacious and murderous attack on people and property at the very heart of the United States financial and military machine?

Or have there been other significant signs, gradual in their nature, that we have barely noticed but which have been changing our lives and the lives of those around us?

I put it to you that a mix of evolving events have been changing the world for some time past. Some of these events have been more noticeable than others. Some have been barely perceptible.

Technology, for example – it's well known that the rate of advance and registration of new patents continues to accelerate at a rate that would have been unthinkable 100 years ago, even 50 or 25 years ago.

Did you hear of the remarks of the head of the United States' Patent Office around 1899 that "everything that needs to be invented probably has been invented"?

Communications and the Information Technology Revolution

Computer and mobile phone technology race forward bringing new opportunities not only to the developed but also to the developing world. As we have known throughout history, better communications can be used for good or evil.

Wealth has increased for many: poverty has also increased. Economic data show that the gap between rich and poor seems to be growing wider. Does this not lead to increased instability?

Stock markets have boomed and fallen, and then boomed and fallen again ... but on balance many people are better off than their parents and grandparents were.

The rate of mergers, acquisitions and consolidations over the past 10 years or so outstrips the rate of corporate activity in the 1970s and 1980s. New dangers arrive with newly acquired companies, especially if they come from less familiar parts of the world: Russia, China, India, South East Asia. More change bringing less certainty.

My own former employer, an American tobacco company, had an LBO in 1989 of about $31 billion. At the time it was the biggest in corporate LBO history. But when converted to today's dollar value it would be peanuts. I'll come back later to the implications of M&A activity on your profession, and the opportunity it presents to the greedy and unscrupulous.

Hand in hand with wealth comes greed. Greed and corruption. This touches on your profession. You are in the forefront of dealing with the consequences of greed and corruption.

Greed and corruption show themselves through the organized crime gangs that may control a region or a port, or a city, or even a country. But greed and corruption also show themselves inside corporations. We have all seen this during the recent high profile trials and convictions of senior corporate executives in the USA.

Has there been a major change in our behaviour? Are we less ethical than our parents and grandparents? I pose the question to you because I am unsure of a scientific response. I can only speak from personal experience and conviction.

Many of you may perceive a decline in ethical standards of behaviour. You, more than most within corporations, are expert observers of human behaviour. Has there been a change in ethical standards? Are we more or less ethical than 100 years ago, or even 50 years ago? If you detect a change for the worse, how has that affected your professional role? And what, if anything can you do about it?

There do seem to be certain paradoxes playing themselves out here.

On the one hand we seem to be becoming increasingly a global village. Yet on the other hand so much of our planet seems polarized by religious and cultural differences.

Samuel P. Huntington writes of the "clash of civilizations" between Christendom and Islam. Others ponder on the resurgence of Islam from East to West

Some thinkers on Middle East politics ask the question, "Who won the Second Gulf War? and answer with one word, "Iran" What are the implications for security of a resurgent Iran determined to take what it sees as its rightful place as the regional power in the Middle East ... and (eventually) a nuclear power to boot.

Regional politics inter-change in a flash with international politics.

The only world superpower is resented. Some would say that is inevitable, just as the British at the height of their imperial power were resented; and so too Napoleon; and so too the Spanish; and no doubt Alexander the Great was not welcomed with open arms when he invaded Persia and beyond.

But is the only superpower in danger of over-extending itself? And what are the implications for security specialists like you if the USA's enemies detected weakness, a change of behaviour, or a longing for withdrawal?

There seems to be a reduction, even a lost of community resilience. Is this loss corrosive? Don't weak communities lend themselves to higher rates of crime? What impact does that have on you? And what on earth can you do about?

But the rapid rate of change makes it more difficult than ever to adjust our solutions to new security threats. Why? Because what worked last time may not work the next time? Generals are often accused of fighting new battles based on their experience of their last war. Is that fair, even though it may be true? Is that a charge that could be levelled at you?

Recently, I was part of a NATO conference on Illicit Trafficking of Radioactive and Nuclear Material. I heard a leading thinker in the British defence establishment remind his audience that technology alone cannot be the solution to better security. A corresponding change was also needed in how we think about and perceive problems and responses. And we need to rethink our experience because during a period of stability our greatest asset in problem solving is our experience. But in times of revolutionary change, like the times we are experiencing today, can be a dangerous hindrance. Experience can be excess baggage. Because it worked last time does not mean it will work next time.

2. Threats: Organised Crime and Terrorism

Organised crime and the ever present threat of terrorism are, perhaps, the greatest dangers confronting you today. And they are increasing.

You know already that any goods being traded on the black or grey markets are more than likely linked to organised crime or terrorist groups.

Some definitions may help to clarify my meaning. The criminal gang is in the illicit business purely for the money and its own benefit. The terrorist is looking for political gain, for control. Although depending on the country, situation and culture, the roles can inter-change. Ireland and Afghanistan are two good examples. You probably know more.

The thirst for power and control represents a menace of increasing proportion.

The amount of money being generated by illegal activity is immense.

Where does the money go? I have just come from a conference of international tax planners in London where part of the discussion touched on the problems of money laundering in offshore jurisdictions, especially for some small islands in the West Indies. However, the bulk of illegal money is generated in the wealthy and, some would say, profligate West. The money has to be laundered in large centres like London and New York.

Perhaps we in the rich North are more part of the problem than part of the solution.

Today, public interest is slowly turning to the growing threat from counterfeit products. And it is not the Intellectual Property argument that exercises people. Is health, safely, and terrorism. Over the past three years I have spoken at the Universities of Geneva and Cambridge and the American International Club of Geneva on the topic: The Globalization of Contraband and Counterfeiting – Victimless Crime or Tip of a Criminal Iceberg?

Though some dispute the figures, estimates of the scale of the counterfeit problem alone are about 5–7% of global trade, that is about $500 billion. Turkey and Australia have GDPs of a similar size.

The OECD estimates the figure could be as high as 9% of world trade and has now looking at a new global study over the next couple of years to obtain a more accurate figure. Good Luck to them! It is badly needed.

In 2001 the European Union seized more than 95 million counterfeit articles.

Ten times more than in 1998. By 2002/2003 the number doubled. The trend goes relentlessly upwards.

In the USA alone, the Federal Bureau of Investigation reckons that the economic impact on legitimate businesses of losses due to counterfeiting is about $250 billion annually. The FBI describes counterfeiting as "the crime of the 21st century".

One example from the luxury goods industry: a raid on a Chinese Diaspora counterfeit watch gang in Manhattan found orders and *pro formas* for business valued at $250,000 per day … and potentially about $50 million worth of business on the books for a year. And this was just one family.

The government in Russia is said to forfeit about $1 billion per year in tax revenue due to counterfeit products. Money it can scarcely afford to lose.

A distressing BBC World Television programme just last week showed the extent of the fake pharmaceutical medicines problem in Africa, but made in India and China.

Yet despite the vast amount of illegal funds generated, and the huge damage done to livelihoods, and national not to mention local economies, the majority of consumers mistakenly see counterfeiting as a "victimless" crime. There is a lot of work to be done here to change public attitudes and behaviour.

Focusing on Specific Industries and Product Categories

The picture should give cause for alarm. Take, for example, cigarettes which by value are unquestionably the biggest product category for smugglers and counterfeiters. Some in the tobacco industry estimate that as much as 4% of world consumption could be counterfeit. That is about 200 billion cigarettes a year. Four times the size of the UK (legal) domestic market. No-one is sure how big the total UK market is today.

In the USA to take another example, 10% of the market is estimated to be illegal. That is about $76 billion in sales value – equivalent to half the GDP of Hungary, and about twice the GDP of Slovenia.

A former colleague of mine in the USA reckons that up to one third of all cigarettes in New York State could be illegal – whether grey or counterfeited.

In the UK, which until fairly recently was one of the most orderly and profitable tobacco markets both for manufacturers and government, tax revenue and regulatory control of the market is sliding out of control. According to official statistics anywhere between 21% and 28% of the market is illegal. Loss in tax revenue to the government is estimated at about $5.4 billion per annum.

Of course there is a double hit to the tax payers as governments also increase expenditure on law enforcement, customs officers and police to fight the very problem that they have helped to create.

In France, formerly a country with virtually no illegal cigarette sector, contraband is now growing slowly but steadily. All thanks to heavy tax increases and neighbours who have not followed suit.

Alcohol, though not on the scale of tobacco, has a large and lucrative contraband market. The EU estimates that its member states lose annually over €500 million in tax revenue from smuggled alcohol. In the UK, recent estimates of lost revenue were as high as €340 million.

Turning to pharmaceutical products, the World Health Organization believes that between 5% and 7% of pharmaceutical products world-wide may be counterfeit, with particularly dangerous and alarming implications for public health. In 2002, the Food and Drug Administration in the United States launched 30 separate investigations into pharmaceutical counterfeiting, many of them popular brands.

There is increasing concern among some of the major pharmaceutical manufacturers in Europe about the porosity of borders to the south east, east and north east of the recently enlarged European Union.

Global estimates of losses to the industry are around $12 billion per annum.

Luxury goods, brand name fashion clothing, and shoes are also regularly counterfeited.

Reasons for Growth in the Illegal Trade

The financial incentive is of staggering proportions. A typical example is as follows. $120,000 is all that is required in China to manufacture sufficient cigarettes for a 40 foot container, that is 8.5 million cigarettes, or 425,000 packs of 20. Shipping and "facilitation", in other words bribery, costs are already included in that number.

In the U.K. the recommended retail of price of cigarettes in the premium sector is over $8 at today's rate of exchange.

If these fake cigarettes can be sold at, say, $6 for a pack of 20 the net profit can be as high as $2,430,000 … with little risk.

Proximity

The next logical question is where can these cigarettes be sold profitably? Well, wealthy Europe is surrounded to the south, east and north east by a ring of much poorer countries.

There is sufficient financial incentive to get profitable but illegal goods across porous borders and not very long distances.

The growth of counterfeited products is alarming. So too is the range of products being faked.

Not only cigarettes, alcohol, ladies cosmetics, baby food, coffee and tea, car components, olive oil, shampoo, batteries, clothes, shoes, garbage bags, pasta, golf clubs, motorcycles, and luxury leather goods.

The police who in the past seemed reluctant to draw direct links between terrorist funding and contraband are a little more open now and say that there were links between 9/11 and gangs involved in contraband activity.

Defining the problem and relationship between organised crime and terrorism is not easy for governments. Friends of mine in Washington, DC tell me that Al Qaeda, for example, should not be referred to as a group but rather a loose, multi-national movement. It is estimated to be present in some shape or form in over 60 countries! That could have worrying implications for you.

But the point of terrorism – maximum impact on legitimate economic or civilian targets in Europe, the USA and the Middle East – makes closer cooperation between organised crime and terror groups more not less likely.

This raises some simple but vital question to international companies: are they doing enough? Are they doing all they can to protect their employees abroad, and to protect their brands? Can governments do more? Should governments and companies be working more closely to influence policies and increase security?

As the nature of security problem becomes more complicated and sophisticated so it will demand better and more integrated responses from governments and companies. Effective solutions will require deeper thought and analysis. This is an area where business should work even more closely with government so as to get over the problem of being reactive or defensive. Business people can take advantage of the trust which the public has in their security forces, which they may not have in the business community, to bring resources, knowledge, and expertise the aid of their government.

Crime and terrorism go hand in hand. Once a criminal network is created, the aim is to move as much through that channel as possible to maximise profits. Business knows a lot about criminal networks. Governments know a lot about terrorist threats. Increasingly, business and governments can and should combine their knowledge in a coordinated fashion to counter the threats.

Cyberspace

There is a new frontier into which extremists and criminals are now moving: the sophisticated use of the Internet and cyberspace. These attacks can be launched from a safe distance to neutralise control of essential economic infrastructures using pure IT resources. Computers are vulnerable.

Financial services, power supply, transport, emergency services, food and health are all reliant on computer based equipment which in turn is increasingly susceptible to hacker, attack, viruses and worms, as well as band width clogging and digital traffic jams.

The digital traffic jam caused by the MSBlast worm is often cited as another contributing reason for the dysfunctional US power stations that were unable to balance the load on the 14 April 2003 and caused the largest power-cut in history.

The cascading failure led to the collapse of electric power in the entire North East of the USA and affected other cities like New York, Detroit, and Toronto the UK, Sweden and Denmark as well as Italy and Switzerland had power outages near the same time.

Fundamentalist hacking activity is rising and has been getting more sophisticated over the past three years. Hacking groups from Kashmir, Pakistan, Morocco, Turkey, Chechnya, Saudi Arabia, Kuwait, Indonesia, and Malaysia are collaborating both with each other and fringe of anti-globalist groups based in the West in order to target international and domestic online assets.

Large and small scale businesses, government computer networks as well as home computer owners have all been targeted by organized crime syndicates and radical hacking groups. The identity theft, the financial fraud and business interruption damage exceeds tens of billions of dollars in each category.

Identity theft, 'phishing' scams targeting over 20 major banks in the world and credit card fraud are all rising. They provide cover for licit and illicit organized crime and extremist activists. As much as 25% of organised criminal syndicate activity is perfectly legal, by that I mean it is licit because they also need a safe façade behind which they can carry out their nefarious activities without challenge.

From spam to 'malware' proliferation, the use of home computer zombies is growing. According to a friend of mine who created a company that now protects some of the world's major banks and defence establishments, every single computer on the planet can be recruited for malevolent crime and extremist activities either as an end target or a go-between for launching Distributed Denial of Service attacks followed by extortion or ransom demands. A number of companies have already paid up. 11.5 million zombies are used for illegal file sharing and mail relays. And the cost of digital crime is in excess of $250 billion per year.

This is a modern, almost post-modern phenomenon. Perhaps this type of security threat has already hit your own company. Maybe it is something that some of you are not comfortable discussing in a wider circle like this. Because the security of corporate information is so important, and are the consequences of the latest technological innovation, it is vital – I would suggest to you – that you position yourselves at the forefront of new ideas in your respective companies to manage and thwart the growing problem.

I shall come to some thoughts of my own on this at the end.

3. Corruption Inside the Firm

Now I turn to my third point: corruption inside the firm, and implications for your profession.

There was a time when your corporate colleagues, Executive VP Marketing, Executive VP Business Development, Executive VP Finance, Executive VP Sales, and even the CEO had more modest ambitions to just run the business profitably, provide the shareholders with consistent dividends, and take home good bonuses at the end of each year.

Since the frenzy of M&A activity that has changed. Many of your corporate executive colleagues have now made great advances thanks to mergers and acquisitions. Their interests, I would suggest, may be much more short term than before. And their interests are much more for PERSONAL GAIN because of the large amounts of money that can be made.

This is the point at which you come right up against personal greed in the higher echelons of management.

Because the M&A market is becoming more and more competitive as companies search for new targets to acquire, special schemes, perhaps even favours, will have to be created to gain advantage or to win contracts.

You are expected to protect the company from attack a number of different sources. And yet my guess is that it may be difficult for some of you to break into the very top end of the senior management circle before it is too late.

Perhaps you can only deal indirectly with the potential problem of senior management who in their efforts to maximize profits, or strike the big deal or acquisition may also be creating liabilities for the company, the shareholders and the employees.

There is yet another dimension which many complicate matters further. When your company is about to embark on a major relationship with, or intends to buy, another company, are you involved in the due diligence? Does the senior management come to you before the fact or after the fact? Perhaps there are awkward conflicts of interest at the top that should be brought out, but for various reasons are suppressed. Are they aware, as you are, that there is more to due diligence than the balance sheet, P&L, Return on Investment, and other important ratios?

As you know, organized crime is growing increasingly adept at putting their place men in the higher echelons of corporations.

A company can be made extremely vulnerable if it does not know what kind of foreign networks it is dealing with, and whether it is unwittingly grafting onto its healthy body a malignancy with organised crime or terrorist connections. Checks by due diligence audit firms will not find out that information so easily. It is important that people with your skills, knowledge, and experience are brought in at the beginning, and are working at the very highest level within the company. It is a question of knowing. You have the knowledge.

So What Could a Better Future Look Like?

- It is clear that information and the technology to use and move it will accelerate.
- It is clear also that governments will seek to try to draw more and more power under their control to protect citizens ... whether efficiently or not is another matter.
- There will not be a reduction in terrorist activity in the foreseeable future.
- Certain foreign nationals and their property, in particular Americans, British, Australians, and some others will be seen, perversely, as more legitimate terrorist targets than others.
- Businesses will come under increasing pressure as oil prices climb, and margins are squeezed.
- You will probably be expected to do more with less.
- The criminal mind will continue to be ever more resourceful dealing in businesses that brings the greatest returns for the least risk.

How do you deal with this? Perhaps a change of mentality is needed. Just like, for example, NATO which is thinking more deeply than ever before about the nature of its response to the security threat to its members.

Through your professional association, like this one – ISMA – you should keep asking yourselves the searching question: what policy in government have we influenced for the better security of our colleagues and friends in the company today? If the answer to your own question is unsatisfactory, what are you doing to improve matters?

Though governments have the elected authority and mandate, you in private enterprise have a sharper motivation, are closer to the problems, and have the variety of networks to provide solutions. Sometimes this may mean collaborating with the authorities, sometimes the problem can be solved by acting unilaterally.

Use your resources and insights to help bureaucracies to be sharper and more focused and more relevant. They are a lumbering machine. You are more nimble and responsive.

Broaden and deepen the thought processes you bring to understanding and solving problems. If you have not already, build links with the best think-tanks and universities to have access to the best brains and original thinkers on and around issues that are vital to you. I mean not only in a forensic sense, but also technology, social and psychological sciences, risk analysis, defence establishments, and the best business schools.

Engage with the media. They are crucial improved understanding on both sides. They have a wealth of information but often not a deep understanding of issues. But they understand how to communicate. And they have access and an audience.

Initiate and encourage regular cross-functional exchange in your corporations at the highest level. But bring a new dimension to your colleagues by raising the bar of knowledge and intellectual challenge, thus forcing them to admit that they had never looked at a particular problem or issue quite like that before until you had the insight and foresight to show it to them.

And finally, show your management colleagues audits that have teeth, and that reveal uncomfortable truths. This will serve two purposes: it will keep management on its toes if anyone is being tempted to do something they shouldn't. And it will keep the company in fitter shape to deal with new threats as they arise.

These have been my thoughts, ladies and gentlemen. I do hope that you have found them as interesting to listen to as I did in preparing them. Thank you again for your kind invitation today, and for receiving me so graciously into your group.

APPENDIX I

1 SEARCHING FOR HIDDEN RADIOACTIVE SOURCES – EXPERIENCE FROM EXERCISES IN POLAND

Genowefa Smagala
Project Leader, Central Laboratory for Radiological Protection, CLOR, Warsaw, Poland

Introduction

Licensing, accounting and national control of radioactive sources dates back to the mid 1960s in Poland. Implementation of additional measures for controlling movement of radioactive sources both across the borders and in the territory of Poland commenced in 1990. National efforts aimed at interdicting an illegal movement of radioactive materials and responding to radiological emergency events involving also orphan sources resulted in a deployment or an modernization of the equipment for detecting and searching such materials. Although the task to reduce the threat to the public and the environment against dangers arising from innocent or malevolent use of radioactive sources was undertaken in the capacity of the country's own needs and resources it has been accomplished with the aid of international and bilateral assistance in the frame of support programs. Assistance provided in the form of training, equipment and joint exercises has contributed to the experience the country acquired in searching for hidden radioactive sources and responding to emergencies with involved radioactive sources of unknown origin. Two exercises, the Turawa 2003 exercise and the Bobrowniki 2004 hands-on workshop, organized by the Central Laboratory for Radiological Protection, CLOR, in Poland were of special importance to the experience the country achieved.

Turawa 2003 Exercise

In September 2003 a regional intercomparison exercise of Mobile Spectrometric Laboratories was held in Turawa in Poland. A donor country, Denmark, and five recipient countries (Estonia, Latvia, Lithuania, Poland and Russia) tested their experience in using Mobile Spectrometric Laboratories, *inter alia* for searching hidden non-shielded radioactive sources. The Mobile Spectrometric Laboratories, delivered to the recipient countries a few years earlier in the frame of assistance program, were equipped with the high volume scintillation detector – NAI(TI) crystal – mounted on the roof of the car, computerized spectrum analyzer on the board, portable gamma detector, hand-held nuclide identifier and Geographic Positioning System. The equipment has been operated and tested by the participating teams in similar exercises before.

The tasks of the Turawa exercise included estimation of distance to the hidden non-shielded radioactive source, identification of radioactive isotope and measurement of its activity. For the purpose of the exercise a Polish radiography company prepared and delivered four radioactive sources and they were: cesium-137 (Cs-137), cobalt-60 (Co-60), iridium-192 (Ir-192) and selenium-75 (Se-75). Every team performed the measurements of the same source from the same distance. The four hidden sources were detected by all teams without a major problem, although the estimated distances to the hidden sources varied significantly between the participating teams. The next tasks i.e. identification of isotope and estimation of activity of the hidden sources were causing more difficulties. Although, no team had a problem with identifying Cs-137 and Co-60, some had a little problem with identifying Ir-192 and most of the teams had a problem with identifying Se-75. There were also differences between the results of activity estimation, likely caused by a different formula used for activity calculation of the hidden sources.

The exercise showed that experience in searching, locating and identifying unknown radioactive sources should be shared also among the practitioners, who operate advanced technical equipment, which can be used for a search and identification of orphan sources. Practical exercise with the use of real radioactive material helped to check capabilities of the participating teams, identify errors and hopefully avoid them in the future, or in tests or in real cases. The results of the Turawa exercise and the problems faced even by the highly trained specialists should be taken into account by responders and decision makers given that a similar technology, equipment with sodium-iodide (NAI) detectors, can be used or is used by the border controlling staff, who is much less experienced in dosimetry measurements.

Bobrowniki 2004 Hands-on Workshop

Another exercise contributing to the experience acquired in searches for the hidden radioactive sources and response to emergencies with involved radioactive materials of unknown origin was the Bobrowniki 2004 hands-on workshop held on 28 September 2004, at the border crossing with Belarus and its surrounding. The exercise was a requirement of the European Commission (EC) support program for the response system to illicit trafficking involving nuclear and radioactive materials in Poland. The national handbook for the response to illicit cases involving nuclear and radioactive material (RITNUM), elaborated by CLOR in the frame of the project launched by the Institute for Transuranium Elements (ITU) of the Joint Research Centre of the EC, was to be validated in a demonstration exercise in the field.

The Bobrowniki hands-on workshop tested collaboration and competence of various authorities and services in case of detection of radioactive and nuclear materials of unknown origin. The exercise was held with the use of samples of real nuclear and radioactive materials taken over in a similar incident in the past. The scenario provided for a possibility of combining radioactive materials with explosives or explosive devices as well as with so-called "blocking of the object" by a group of criminals.

The exercise included two interrelated cases:

- In Bobrowniki border crossing with involved a seizure of radioactive and nuclear material
- At an abandoned farm beyond the borderland zone, where other such materials and explosive devices were stored and guarded by the armed people

And it was continued at the laboratories of the CLOR and of the Police to perform analysis of the seized and collected materials and to carry out investigations of the evidence.

On 28 September 2004 at 5 a.m. an officer on duty of the Border Guards in the Bobrowniki border checkpoint stopped a car when his pager signaled increased level of radiation, after information about a planned illicit trafficking transfer of nuclear and radioactive materials through the borders of Poland passed on to all border services at the Polish Eastern Border. The key players were the services and authorities of the Podlasie Province, respectively to the accepted scenario of the exercise and involved inter alia Border Guard, Customs, Police, Sanitary-Epidemiological Station, Fire Brigade, medical ambulance. The practicing services were joined and supported by CLOR to carry out radiometric surveillance, contamination assessment, searches for hidden radioactive sources and categorization of the seized substances. In all the places of the exercise i.e. at the border crossing, in the abandoned farm and in the laboratories afterwards, action was in line with the requirements of the Law and provisions of the RITNUM Handbook.

The exercise was observed by a group of over 100 persons representing 31 national institutions, 6 international organisations or foreign countries (EC, IAEA, EUROPOL, ITWG – Nuclear Smuggling International Technical Working Group, Belarus and USA), as well as by a group of journalists from public mass media from Podlasie District of Poland. Action at the incident scenes was commented for observers by representatives of the practicing entities. The very first comments and opinions on the exercise were shared and conclusions drawn at the summary session after terminating the in-field actions.

The categorization of nuclear and radioactive materials during the exercise was carried out with the use of hand-held nuclide identifiers, based on scintillating detectors, whose resolution is insufficient to identify a mixture of radioactive or nuclear materials. The samples of nuclear and radioactive materials used in the Bobrowniki exercise (low enriched uranium pellets, depleted uranium powder and cobalt-60 plates), as well as collected environmental samples were taken to CLOR for detailed analysis. Examination of the samples at CLOR was partially performed with the use of upgraded, under the EC project, high resolution gamma spectrometry equipment. After the CLOR laboratory investigations samples of uranium pellets and powder were sent to the ITU for a joint, with the CLOR specialists participation, nuclear forensic analysis. The ITU nuclear forensic analysis scheme included characterization of the materials, determination of a potential producer and interpretation of the findings.

The exercise contributed to the enhancement of co-operation among all the involved authorities and services. Competence was checked and experience shared not only among the participating services, but also between the donor and the recipient nuclear forensic laboratory. Analysis of the exercise, the gained experience and the revealed gaps were used for the system improvements so that similar situations can be avoided in the future.

Conclusions

- The in-field exercises with the use of real material were a great experience for the all participants.
- Experience in searching, locating and identifying of hidden or abandoned radioactive sources should be shared also among the practitioners and highly trained specialists.
- Only a preliminary categorization of nuclear and radioactive material can be performed in the field with the use of hand-held nuclide identifiers, based on scintillating detectors.
- Systematic training, sharing experience in practice and close cooperation between all institutions involved in prevention, detection and response systems significantly contribute to the fight against terrorism.
- International and bilateral support programs give good results only when both sides, the donor and the recipient, are willing to cooperate.

2 LETTER FROM US STATE DEPARTMENT IN SUPPORT OF THE WORKSHOP INITIATIVE

United States Department of State

*Under Secretary of State
for Arms Control and International
Security*

Washington, D.C. 20520

September 10, 2004

Mr. John W. Wood
Dr Evgeny Velikhov
Directors
NATO Workshop: Illicit Trafficking - Detection,
   Deterrence & Interdiction

Dear John and Evgeny:

Congratulations to the Trilateral Group and NATO for putting together this timely workshop to engage in an in-depth review of mechanisms to detect, deter, and intercept the trade in weapons of mass destruction.

The policy of the United States has been to make effective and robust use of our international and national legal authorities, regimes, and alliances, to address the major challenge to peace and security posed by the proliferation of weapons of mass destruction. UN Security Council Resolution 1540, initiated by President Bush and unanimously adopted by the Council, "calls upon all states to ... take cooperative action to prevent illicit trafficking in nuclear, chemical, and biological weapons, their means of delivery, and related weapons." We are working on a variety of fronts to meet this challenge.

Our most prominent initiative is the Proliferation Security Initiative (PSI), which involves the high level political commitment of participant nations to take robust action to interdict suspect shipments. By working together to make maximum use of national assets and capabilities, consistent with national and international legal authorities, PSI participants are interdicting illicit shipments and denying traffickers and other facilitators the means to trade in nuclear, chemical or biological weapons, their means of delivery and related materials. Through the PSI, like-minded countries are training together on land, air and sea, working with industry, reviewing their respective legal authorities, and sharing information and lessons learned, so that when timely information presents itself, nations have the capacity and the will to respond quickly and effectively.

Complementary to the PSI, at NATO we are working with our treaty partners to improve the ability of our NATO Alliance to address more actively the threats posed by weapons of mass destruction. While Operation Active Endeavor is directly targeted at interdicting maritime vessels carrying terrorists and items of support for terrorism, it presents opportunities for NATO to engage more aggressively to halt the traffic in weapons of mass destruction by terrorists in its area of operation. This concept has been under consideration at NATO, and we encourage NATO to continue to think creatively about how it might appropriately include WMD interdiction efforts as part of its mission. Similarly, we are encouraging regional leaders to work with their neighbors to develop such capabilities. For example, Japan is playing a leadership role in Asia and is inviting other ASEAN nations to participate in a maritime exercise it is hosting next month; Poland has been playing a leadership role with its neighbors in Europe; and Portugal has engaged with countries in Africa.

The threat posed by an unhindered trade in weapons of mass destruction is clear. For the first time the UN Security Council stated explicitly that such proliferation is a threat to international peace and security. How we respond to this threat will set the stage for the kind of world we will leave for future generations. I know it will be a successful and insightful three days. I look forward to hearing about the recommendations proposed by your distinguished participants.

Sincerely,

John R. Bolton

3 LETTER FROM BARONESS THATCHER IN SUPPORT OF THE WORKSHOP INITIATIVE

Margaret Thatcher

3rd November 2003

Dear John

When we dined the other week you mentioned the prospect of my taking part in the follow up to the inaugural conference on International Approaches to Nuclear and Radiological Security, which you Co-Chaired last year with Dr. Evgeny Velikhov.

This I am delighted to do, and I am also delighted that you and Dr. Velikhov will once again serve as the Chairmen. In my view there is no more important work. Indeed, as I said in my speech last year, "we are now forging a partnership which will bring dividends both for us and for future generations...You serve not only your own nations, but all nations, by your efforts."

I must congratulate you for the central role which you and your colleagues continue to play in encouraging and keeping this vital community together, and also Secretary Abraham for his dedication to this great work. I very much look forward to meeting him here in London at next year's conference.

Your sincerely

Margaret Thatcher

Mr John W Wood

THE RT. HON. THE BARONESS THATCHER, L.G., O.M., F.R.S.
HOUSE OF LORDS, LONDON SW1A 0PW

APPENDIX II

**BACKGROUND PAPERS FROM
INTERNATIONAL APPROACHES TO
NUCLEAR AND RADIOLOGICAL SECURITY
CONFERENCE 2002 (IANRS)**

1 CHAIRMEN'S SUMMARY AND EVALUATION

John W. Wood and Dr. Evgeny Velikhov
Conference Co-Chairmen

Dear Colleagues,

This inaugural conference has considered what is arguably the most pressing question facing the international community today. What are the threats from the potential proliferation of nuclear and radiological weapons, and what are the most effective processes and means by which the threats can be contained?

The strategic realities which now confront us have changed beyond all measure from those which confronted us a mere decade ago. Then, nuclear capabilities were confined to states alone and disciplined by a menacing, but well calibrated, international regime. Now, nuclear and radiological capabilities are wider spread, easier to come by and more difficult to control. Then, deterrence worked because nuclear capabilities were possessed by states with definable borders and stable interests to preserve. Now, the worry is that deterrence does not work because nuclear capabilities are sought by loose coalitions of non-state actors operating globally, who have no interests to preserve, but only values to proclaim and, often, those values are the nihilist ones of glory by self-destruction. Then, many thought the nuclear age was coming to an end. Now, it is apparent that it is only just beginning. At a state level, it is true, the paradigm of deterrence is alive and well. But this fact can, in itself, produce unforeseen consequences, for the more that deterrence can be seen to have worked for great powers who had a vested interest in mature stability, so that lesson has been also been absorbed by a new tier of lesser powers, many of them either unstable in themselves, or located in unstable regions.

It is a melancholy truth that the right lesson appears to have been learned by the wrong people for the wrong reasons: that the possession of nuclear weapons, whatever their apparent purpose, seems to confer a legitimacy upon the state which might choose to embed a nuclear emblem in their flag: legitimacy, in the sense of being capable of both independent decision, and destruction. As the twentieth century great power competition demonstrated, that deterrence does work, so in the twenty-first century, increasingly unstable nations might be tempted to wear the nuclear badge in direct proportion to the degree to which they feel threatened by the international system. The possession of nuclear capabilities can become a paranoid's security blanket. The implications of these considerations are alarming in the extreme.

The worst possible outcome would be that nuclear and radiological materials fall into the hands of those who actively seek to use them as weapons, rather than as instruments for the containment of weapons. Or that others seek to acquire them either to make their regimes inviolable to change and influence by the international system, or as means of coercing other states. In the last half of the twentieth century the system of deterrence gradually changed from being an instrument for the avoidance of war to becoming a framework for the containment of wars. Then, deterrence was institutionalised. Now, in the beginning of the twenty-first century, we are confronted with an utterly changed situation. Deterrence is de-institutionalised. We find that many of its assumptions are irrelevant in the new world of globally spread, loosely aligned, fanatically committed cells rather than states, and formerly non-nuclear states, either seeking to prevent internal change or to establish local, regional and international abilities to coerce or threaten.

It is against this background of the diffusion of nuclear and radiological science on the one hand, and increasingly fragmented domestic societies and a diverse international system on the other, that this conference was convened. Our task simply was to begin to rethink strategic thinking in the new context of the threat of nuclear and radiological proliferation. To analyse, find and agree ways of dealing with the increasing threat of the proliferation of nuclear and radiological materials. Given the high probability that proliferation cannot be prevented, our task was to find the most effective and vigorous possible measures for its containment. The job of our deliberations was to act as a catalyst for organising the international and nuclear and radiological community in such a way that cooperation would be facilitated, gaps in security and defences identified, and the beginnings of an effective and agreed international regime coordinated. To bring peace under control, few callings can be higher.

A hallmark of this conference was the sustained diligence of participants in working through the issues which confronted them and the high level of cooperation and agreement achieved. The welcome outcome, and one which is rare, was the production by each Working Group of actionable recommendations. These results are a testimony not only to the seriousness of purpose brought by all attendees, but particularly a tribute to the quality of sustained commitment over a four day period, of the working group chairmen. For this reason we would like to pay a particular tribute to Professor Leonid Bolshov who chaired the Radiological and Threat Reduction working group, Mr. Kenji Murakami who chaired the Trends in Illicit Trafficking group, Dr. Jack Caravelli who chaired the Bilateral and Multilateral Approaches to Material Protection, Control and Accounting group and to Ambassador William Courtney for chairing the Emerging Challenges to Preventing Nuclear and Radiological Terrorism group. These Chairmen evidenced both intellectual distinction and leadership abilities of a high order. A tribute should be paid to the speakers in each session of the working groups. Their knowledge, expertise and hard work provided the technical and intellectual engines for each session. But in the end, the quality of the output of the working groups was determined principally by the involvement, responses and contributions of all the attendees. And it was in this dimension that this inaugural conference was perhaps most distinctive. The extraordinarily high quality of all attendees was equally matched by their willingness to contribute.

The conference's plenary speakers, and guest speakers, from their different perspectives, created not only challenging and stimulating professional context for the conference, but also set the overall intellectual framework within which our deliberations took place. From the opening address by Ambassador Linton Brooks we were challenged to determine what role the international community would play in dealing with the issues at hand. Undersecretary of State John Bolton, in his keynote address, laid out in stark form the sharply changed international context within which, since September 11, 2001, issues of nuclear and radiological security could be considered. And Deputy Minister Kotelnikov underscored the need for the international community swiftly to find ways of coordinating its international anti-terrorist activities. Speaking from the

European perspective, Mr. Paul Schulte presented a lucid and challenging exposition of the intellectual and practical challenges created by the new international situation. Finally Mr. Tsukasa Hirota presented a comprehensive view of the issues as seen from the Japanese perspective. The speakers of the second plenary maintained the high standard begun by the first. Ms. Thérèse Delpech analysed the new strategic background against which our deliberations were taking place, argued the need for Europe to increase its participation in cooperative endeavours, and laid out clear priorities for action. Secondly, The Right Honourable Bruce George MP presented a wide-ranging *tour d'horizon* of the threats and the political practicalities of dealing with them, and underlined the need for generating public confidence in nuclear and radiological security policies. Finally, Ambassador Kenneth Brill contributed a wise and lucid case for an integrated approach to nuclear material security and outlined six elements which that approach should contain.

In the third plenary, Mr. Hiroyshi Kurihara set the deliberations in context by combining a history of the development of multinational and national systems of physical protection of facilities, and the roles of multinational and national physical protection systems, with a strong case for the strengthening of the international regime to meet social and political changes in a post cold war era. Following this, Dr. Oles Lomacky argued for the reinforcement of the systems of creating non-proliferation through international scientific cooperation.

Taken together the contributions of the plenary speakers provided a powerful, intellectual framework against which a new and effective regime of nuclear and radiological security could be set.

Because of the complexity and seriousness of the subject, we sought to add a further level of expertise by including talks by speakers from related disciplines. At the Speakers' Dinner we were honoured to hear from Dr. Nikolay Platé, Vice President of the Russian Academy of Sciences, and Professor Lev Sandakhchiev, Director General of the State Research Centre Virology and Technology (Vector) of the Russian Federation. These distinguished Academicians sought to broaden the scope of our deliberations, from the purely nuclear and radiological, to related fields of WMD. As an indication of the high level of Russian support for the conference, Dr. Platé noted that of the special group of ten senior Academicians charged by President Putin with working with the international community in developing comprehensive anti-terrorist approaches, six of them were in attendance at the conference. Professor Sandakhchiev drew attention to the fact that the principal technical solutions which can ensure the safe handling of nuclear and radiological materials can be successfully applied in both the chemical and biological spheres. The presentations of these distinguished speakers clearly set a future agenda of integrating approaches from the nuclear and radiological to the chemical and biological realms.

At lunch on the second day, Congressman Curt Weldon analysed the effects of, and underlined the challenges posed by, the new context created by September 11. He indicated the scale and quality of the response of the US government to these challenges and reinforced from deep experience the absolute need of agreeing modalities by which the international community can work together in transparency and cooperation to meet the new threats. He paid particular attention to the benefits of learning from the successful cooperative experiences of the Russian and American governments in improving nuclear and radiological security.

Lord Powell of Bayswater, our second luncheon speaker, outlined the virtues of taking a bold approach to recognising the new problems confronting the international community and diligently cooperating to devise effective means for dealing with them. While not shrinking from the difficulties involved, and fully understanding that differing nations, even cooperating ones, can have different perceptions and interests, he reinforced the view that first, cooperation can be effective and second, that it can't be avoided. Finally, he cited the inaugural conference as a model of how international cooperation ought to be conducted.

The speaker at the final plenary, The Lord Guthrie of Cragiebank, presented an overview based on deep practical experience of some of the pressing nuclear threat and proliferation issues ranging from those on the Indian Sub-Continent to those in the Middle East, to the global ones confronting the international community as a result of the capabilities and predilections of modern international terrorism. This profound analysis based upon the hard-earned experience of a serving officer and former Chief of the Defence Staff, took us from the technical to the military to the political and provided the conference with some sense of the practical issues confronting military thinkers when confronted by the more technical problems which we had been deliberating.

On the final day, each working group reported its findings and recommendations. Working Group A, **Radiological Threat Reduction** set both long and short term goals. The short term goals were to focus on radiological sources of immediate concern, while long term goals were to create mechanisms by which states could cooperate to regulate and control radiological materials within their own borders based upon internationally agreed standards. This group highlighted a number of issues which would have to be confronted to meet these goals. These included agreeing clearly defined goals, dealing with questions of public perception and trust, helping with the lack of adequate accounting of radiological materials, and questions concerning the adequacy of funding of security efforts and training and assistance of budgets. It proposed actions in technical implementational and political spheres and sets standards by which the results of considered action could be judged.

Working Group B, **Trends in Illicit Trafficking**, conducted a comprehensive look at overall trends, and from this looked at enhancing the cooperative responses. It then judged these against an analysis of comparative program responses. Finally, it looked at, and developed, recommendations for future responses and preventative techniques. This group highlighted the role of international organisations as being essential for coordinating the vital work, examined the constraints upon it and pushed for an appropriate degree of integration of effort. It recommended the setting of immediate targets, the institution of the systematic approach for regular assessment and monitoring and then, strongly recommended a follow-on conference together with the development of a mechanism for close communications and information sharing among participants.

Working Group C, which dealt with **Bilateral and Multilateral approaches to MPC&A** highlighted the fact that any cooperative program in this field will, of necessity, be a technical effort operating in a political world and therefore would need to have appropriate expectations. It argued that the most difficult issues would likely to be political rather than technical, or even financial. It underlined the importance of establishing guidelines to ensure uniform and technically sound security improvements. It stressed the absolute importance of *sustainability* for the effectiveness of any bilateral or multilateral MPC&A in up to date programs. This group looked carefully at existing constraints, focused on the necessity for developing legal and regulatory frameworks and argued for leveraging existing programs and frameworks and to create improvements on a global basis. Finally, it concurred that a follow-on conference had a key role to play in continuing the integration of the international community begun at the inaugural conference. It underscored the utility of creating a communications mechanism for all involved so that the momentum could be maintained and accelerated by a steady flow of personal communications.

Working Group D focused on four key aspects of the **Emerging Challenges to Preventing Nuclear and Radiological Terrorism**. Having considered consecutive responses to new threats, the impact of globalisation on physical protection policies and practices, the development of multilateral approaches, and the creation of focused response and prevention looking towards the future, the group made a series of well thought out recommendations. These included high priority efforts to prevent nuclear and radiological terrorism by all states, increasing cooperation with Russia and other states of the FSU, the implementation of the G8 integrated and coordinated approach, the global partnership program and, in so far as the security and research reactors required to be upgraded, made recommendations for better layered protection in lessening the risks and consequences of terrorist use of nuclear and radiological materials. This group made 16 actionable suggestions, including: one touching upon public education and debate; analysis of laws and enforcement systems; the power of the private sector; the need to analyse and quantify risks; utilising the NPT treaty as a cornerstone of a nuclear non-proliferation regime; and finally the creation of an Internet-based information communication system to facilitate cooperation and communication and to be used as a channel of information sharing and notifications. The group envisaged that all participants would contribute analysis as well as content for the databases and underlined the fact that the system would have to have security mechanisms.

In summary, the conference provided an auspicious beginning: a multi-disciplinary gathering of experts from 26 countries in which debate was vigorous and rigorous, and yet in which agreement was almost unanimous. The extraordinarily high level of contributions across the board were a reflection on the uniformly high quality of attendees. A consensus emerged as to what should be done, who should do it, and in what order.

It is rare for a conference of this scale and scope to produce a concrete set of actionable recommendations. Yet that was a notable outcome. It is even more rare for a conference of four days' duration to maintain the same level of energy and enthusiasm from beginning to end. This was a tribute to the dedication of all concerned and a testimony to the shared appreciation of the importance of the problems under consideration.

Finally, the most important outcome of all was noted by a speaker at the closing plenary that: "What has happened here is that a new community has been brought into being. A community dedicated to finding solutions to the most pressing problems of our time." In short, the overall achievement was to provide an eloquent answer to Ambassador Brooks' opening question: "How will the international community respond?" By wholehearted, determined and cooperative action.

Lady Thatcher in the speech which was, by common consensus, the highlight of the conference, put the point with characteristic clarity, force and foresight. "We are now forging a partnership which will bring dividends both for us and for future generations. You serve not only your nations, but all nations by your efforts. Because though we might wish for the best, we must prepare for the worst, and that is why your conference could not have come at a more crucial time."

2 FINDINGS OF THE WORKING GROUP ON TRENDS IN ILLICIT TRAFFICKING

From the Proceedings of the International Approaches to Nuclear and Radiological Security (IANRS) Conference, 2002 – London, UK
Hosted By: The United States Department of Energy, National Nuclear Security Administration, Office of International Material Protection and Cooperation
Co-Hosts: The Ministry of the Russian Federation for Atomic Energy (MinAtom) and The Russian Research Center, Kurchatov Institute

Introduction

The IANRS initiative launched in 2002 created an international community of experts from 27 countries organized in seven working groups to accelerate and expand international programs to secure nuclear and radiological materials and prevent nuclear terrorism. The inaugural IANRS conference examined how the events of 9/11 added complexity to the problem of preventing nuclear proliferation, securing nuclear and radiological material, and preventing nuclear terrorism, and sought to respond to the international community's call to develop a global response. Accordingly, the principal focus of the conference was developing international approaches to accelerating and expanding programs that secure nuclear and radiological materials. The working groups discussed strengths and vulnerabilities in the current programs and systems and developed strategies to make physical security of nuclear material comprehensive.

IANRS Working Group B, Trends in Illicit Trafficking

Session 1 – The Role & Vulnerabilities of Transit & Border States

Issues
- Current situation in illicit trafficking cases in transit and border States
- Gaps and problem areas in detection of and response to illicit trafficking
- Improvements needed for the customs, border controls and police forces in training and detection/identification instruments

Session 1 – The Role & Vulnerabilities of Transit & Border States

Summary of Discussions
- Four hundred and three confirmed incidents (IAEA database: 43% involved NM and 57% involved RS). Frequency of incidents was constant for NM and has significantly increased for RS.
- Cases so far may be the tip of the iceberg, we don't know what we don't see – Don't be complacent.
- Most seizures were by intelligence information but majority of the cases in Russia in 2001 were by monitoring systems.
- State of the art border monitoring equipment, portal monitors, radiation pagers and multi-purpose hand-held identifiers are of vital importance.
- Customs and border police must have a close relationship with all national-level organizations to enhance targeting.
- Commercial trade must be encouraged to adopt security measures in exchange for expediting at borders.

Summary of Recommendations
- Information sharing among customs and border security organizations should be enhanced.
- Monitoring equipment must be compatible with the level of users and should be supplied with follow-up performance checks and maintenance programs.

- Training programs should be organized jointly with relevant authorities, (police force, customs, and border controls).
- Regional training and seminars need to be held regularly to ensure effective coordination of across borders.

Session 2 – Enhancing Cooperative Responses to Illicit Trafficking

Issues

- Achievement, obstacles and difficulties experienced in the international or cross-border cooperation
- Effective mechanism of regional and international cooperation (role of international organizations)
- International guidelines – establishment and implementation

Summary of Discussions

- Cross-technical training and information sharing with border officials and law enforcement officials are critical.
- The obstacles experienced in cooperative responses include bureaucracy and possible corruption within organizations.
- Successful cases of cross-border training between neighbouring Eastern European states were presented as example.
- Need for cultural change of law enforcement officials and enhanced criminal investigation trainings.
- Idea of "model border crossings" would be useful for training and improving cooperation.

Summary of Recommendations

- The international guidelines (IAEA TECDOCs) should be widely implemented.
- Training courses need to be enhanced by developing a standard syllabus for customs and police officials.
- Training courses should be aimed at training-the-trainers and carried out on a regional basis.

- It is desirable to develop an international clearing-house of different databases from various organizations.

Session 3 – Comparative Programmatic Responses

Issues

- Lessons learned and difficulties experienced (training, technology, information) in programmatic responses
- Application of lessons learned into other regions
- Optimal installation of necessary equipment and needed information

Summary of Discussions

- To reduce risk of illicit trafficking we need preventative measures, efficient intelligence networks and advanced technical means ensuring sufficient sensitivity of detection without restricting flow of goods and people.
- International, multilateral, and bilateral coordination will improve efficiency of deployments and optimization of resources.
- Gaps still exist in technology to cover all ranges of detection; need better hand-held monitors and improvements in portal monitor system.
- There is a need for tighter security control at seaports and installation of monitors of container handling equipment.

Summary of Recommendations

- The lessons learned should be shared between regions and countries by iterating this information into the IAEA TECDOC on border monitoring.
- Special applications (crane and container handling equipment) of detection systems need to be tested and if found practical implemented.
- Detection equipment for both gamma and neutron radiation should be more widely utilized and should be supplemented with specialized investigative training.

Session 4 – Responses & Prevention – Looking to the Future

Issues

- Coordination mechanisms between donors and recipients (role of international organizations)
- Objective and target of achievement in the immediate future and long-term future
- Integration of different measures of nuclear security (illicit trafficking, physical protection, safety and safeguards)
- Next steps and follow-ups of this Conference

Summary of Discussions

- The role of the international organizations is essential for effective coordination as well as clear commitment and willingness by donors and recipients.
- Constraints still exist in collecting incidents information: lack of reporting and completeness, confidentiality limitation and accuracy of data.
- Preferable to achieve an appropriate degree of integration at national level of different measures (Illicit trafficking, PP, Safety and Safeguards) by establishing a single Authority to coordinate relevant activities.
- Essential to set up a well defined national contact points with the international organizations.

Summary of Recommendations

- Set an immediate target to enhance the national prevention and response capabilities with a long term goal to a regional integration.
- Institute a systematic approach for regular assessment and monitoring of the assistance received and its impact on beneficiary performances.
- The follow-up of the Conference is to maintain close communication, sharing of information among participants, establish specific follow-up actions.
- This type of Conference is very useful and hope it will be repeated.

3 FINDINGS OF THE WORKING GROUP ON RADIOLOGICAL THREAT REDUCTION

From the Proceedings of the International Approaches to Nuclear and Radiological Security (IANRS) Conference, 2002 – London, UK
Hosted By: The United States Department of Energy, National Nuclear Security Administration, Office of International Material Protection and Cooperation
Co-Hosts: The Ministry of the Russian Federation for Atomic Energy (MinAtom) and The Russian Research Center, Kurchatov Institute

Introduction

The IANRS initiative launched in 2002 created an international community of experts from 27 countries organized in seven working groups to accelerate and expand international programs to secure nuclear and radiological materials and prevent nuclear terrorism. The inaugural IANRS conference examined how the events of 9/11 added complexity to the problem of preventing nuclear proliferation, securing nuclear and radiological material, and preventing nuclear terrorism, and sought to respond to the international community's call to develop a global response. Accordingly, the principal focus of the conference was developing international approaches to accelerating and expanding programs that secure nuclear and radiological materials. The working groups discussed strengths and vulnerabilities in the current programs and systems and developed strategies to make physical security of nuclear material comprehensive.

IANRS Working Group A, Radiological Threat Reduction

This set both long and short-term goals. The short term goals were to focus on radiological sources of immediate concern, while long term goals were to create mechanisms by which states could cooperate to regulate and control radiological materials within their own borders based upon internationally agreed standards. This group highlighted a number of issues which would have to be confronted to meet these goals. These included agreeing clearly defined goals, dealing with questions of public perception and trust, helping with the lack of adequate accounting of radiological materials, and questions concerning the adequacy of funding of security efforts and training and assistance of budgets. It proposed actions in technical implementational and political spheres and sets standards by which the results of considered action could be judged.

Summary of Radiological Threat Working Group Recommendations

Presented by Professor Leonid Bolshov, Director IBRAE, Nuclear Safety Institute, Russian Academy of Sciences, Russian Federation

Goals

Short-Term: Focus priorities on radiological sources of immediate concern while assisting nations to develop the infrastructure and capabilities necessary to provide adequate materials security and emergency response.

Long-Term: All states should regulate and control radiological materials within their own borders, based upon internationally agreed upon standards. Nuclear and radiological materials should be the responsibility of individual countries. Emergency response system in each country should be prepared to respond to the radiological threat. International cooperation should be developed. Public and decision makers should be prepared.

Issues

1. Clearly defined goals must be developed to help focus the efforts:

   - Do we wish to seek a complete elimination of the threat?

   - Are we willing to entertain the thought that the threat might be beyond our capacity to contain and the use of an RDD is inevitable?

- How is failure defined in this effort and what is an 'acceptable' amount of loss or disruption within that definition?

2. Public perception and mistrust of nuclear and radiological materials (Radiophobia):
    - Lack of trust in governments by the public
    - Lack of trust and transparency between governments

3. Lack of adequate accounting of radiological materials in all countries:
    - Need to develop worldwide inventories that can be independently and readily verified.
    - This should be tied to some form of international agreement that provides a mechanism for enforcement.
    - Should materials be leased and returned to specific countries after use rather than sold outright?

4. Address the scope of the problem and where to focus priorities:
    - Thousands of radiological sources lack adequate security or are 'orphaned.'
    - Need to prioritize efforts to provide maximum reduction of the threat.

5. Concern that funding of security efforts must be provided to countries in need of assistance, while allowing them to develop their own individual systems for control and accounting.

6. Develop and train, and assist other countries in developing and training, emergency response procedures and personnel.

ACTIONS

What Should Be the Results of Concerted Action?

- The acquisition of radiological material for use in a weapon should be made more difficult and expensive (to prevent it completely is probably impossible).
- We should plan for failure, that is, we should be prepared to take action to remediate the effects of such a weapon.

- Our plans should include means to minimize the effects of radiological weapons, including the psychological effects based on a commonly-held nuclear phobia.

What Should Be Done to Deal with the Threat?

Technical

- Know the world wide inventory and catalogue it.
- Prioritize the protection and control of radiological materials.
- Set up appropriate protection, control and accounting.

Implementation Against Radiological Weapons

- The ability for timely damage assessment.
- Prioritize damage and remediation (100% remediation is probably unaffordable).
- Learn to live with imperfection – it might take several incidents with huge costs before a common-sense and practical solution takes place.

Political

- Need to have worldwide standards for determining who (or what country) may be permitted to have or use radioactive materials. (Does a country have the political stability, infrastructure, or leadership appropriate for use of radioactive materials?)
- Need to have adequate laws and sanctions governing the use and storage of such materials.
- Need to develop honest and realistic assessments of the consequences of a radiological attack.
- The populace needs to be educated about the real dangers of radioactive materials so that fears are not exaggerated.

Recommendations on Radiological Terrorism Prevention

1. Monitoring of accessibility of IRS using a comprehensive analysis of all data

2. Development of recommendations and programs for immediate actions to restore adequate IRS control and protection

3. Development of the concept and programs on improvement of national and international systems for monitoring and accountability of IRS, RW, RM

4. Development of scientific and analytical base for justified recommendations on prevention, countermeasures and mitigation:
   - Systematic approach to the area
   - Realistic models
   - Full-scale analysis of different scenarios, including cascade effects
   - Prioritization of possible events on risk base
   - Utilization of practical experience in mitigation of real accidents

5. Adjustment of Emergency Response Systems:
   - Specific response procedures
   - Adequate methods and models for consequences assessment and recommendations for mitigation
   - National special technical support centers
   - International system of communication and technical support

6. Training of first responders and decision makers:
   - Computer systems for training
   - Table-top exercises

7. Adequate perception of radiation risks by population and decision makers:
   - Information
   - Education
   - Consolidation of expert opinions

8. Perfection of legal base in radiation safety:
   - For normal everyday use
   - For emergency

9. Life cycle management of IRS (leasing, terrorist resistant technologies)

4   THE RUSSIAN ACADEMY OF SCIENCES AND INTERNATIONAL COOPERATION

Nikolai Platé
Vice-President, Russian Academy of Sciences, Russian Federation

The IANRS Conference (2002) we attend is an extremely important one. The first day of it has already shown what are the main problems we are faced in the area of Nuclear and Radiological Security and what practical steps should be undertaken in the light of vital challenges the modern world of civilization is meeting when international terrorism is trying to overthrow it.

Our conference is devoted mainly to the nuclear and related problems but I think that the main principles, strategies and general approaches we'll elaborate are absolutely applicable to the chemical and biological antiterrorism policy. This last ones are of the same vital importance as we discuss here now and may be the next conference of this type should be devoted exactly to these two issues.

What have we done in Russia during the last year in strengthening the efforts of political power and of scientific community to get more security on national and international level?

To recognize the role of science and scientists, last year the Presidential Council for Science and High Technology was created under the Chairmanship of President Putin. This council (Academician Velikhov and myself are members of it) together with the National Security Council has considered and approved the main principles of the technological and socio-economic policy of Russia where the role of science and of scientific approaches has been stressed. One of the aspects of this policy is to prevent by all means international terrorist attempts to explode our world and to guarantee the safety and security of the country.

Some of the recommendations of this Council are already taken as practical steps in presidential and governmental initiatives.

The Russian Academy of Sciences is playing an important role in the activity of this Council and we are working in close contact with the National Security Council in this direction.

In summer 2001, three months before September 11, the US National Academy of Sciences together with the Russian Academy organized and realized in Moscow a three-day seminar "Struggle Against High Tech Terrorism in the Modern World" where various aspects of nuclear, chemical, biological, radiological, computer, electromagnetic and other types of terrorism were discussed and some important recommendations how to fight it were made.

A group of prominent American scientists and specialists led by Professor Sig Hecker from Los Alamos (who is with us today) took part in this seminar.

So even before September 11, we understood the danger and probability of terroristic attacks on a global scale. It seemed to the participants of the seminar that we have discussed practically all aspects of technological terrorism. Unfortunately, we underestimated this evil and did not take into consideration the aircraft attack on the skyscrapers.

Now we are preparing the next bilateral seminar with our American colleagues. To activate and to unify the efforts and the experience the specialists of Russia and USA have until now in the antiterrorism policy, the special joint Committee composed of members of both National Academy of Sciences in Washington and Russian Academy of Sciences has been organized this spring. The initiative came from the President of NAS Bruce Alberts and this committee has two co-chairs – Professor Sig Hecker and Academician Evgeny Velikhov.

It is worth mentioning that six out of ten members of this committee on the Russian side are present here at this conference.

The aim of this committee is to analyze different technical approaches how to identify possible terrorists, how to prevent the terrorist attacks and how to minimize the damage if it happens. The main goal is, however, to elaborate proposals and suggestions to the White House and to the Kremlin, as well to corresponding ministries and organizations as to what should be done on national and international level in legislation activity, in technical measures and education and humanitarian activity to fight with terrorism as with one of the global danger for all of us today.

Within the Russian Academy of Sciences we created a few months ago a Special Consultative Council for the problems of fighting against international terrorism.

This Council chaired by the Academy's President Professor Juri Ossipov (the Vice-Chairmen are Vice-President Nikolay Laverov, former Vice-President Vladimir Koudryavtzev and Vice-Chairman of the Security Committee of the State Duma) has three sections.

1. A Section on social, criminological and legal measures to oppose the terrorism
2. A Section on scientific and technical counteractions to terrorism
3. A Section on interaction with national and international public organizations and mass media

Several sessions of this Council have already been organized, the recommendations of our council were submitted to governmental agencies, have been approved and part of them are already in realization.

As one of the results of this activity may I present to our Chairman a brochure which came out of print just last week about social and psychological problems to fight international terrorism. In the beginning of 2003 we plan to make another book about technical aspects what are the problems and how to oppose high-tech terrorism.

Modern terrorism has no religions or national frontiers and that is why beside the technical measures in antiterroristic activity we should seriously think about national and international education programs to teach and to train youngsters how to live in civilized world.

Chemical and Biological Terrorism

A few words about Chemical and Biological Terrorism: they are very dangerous and in some aspects more difficult to fight against these types of terrorism. It is often said that the chemical weapon is the atomic bomb of poor and undeveloped countries. Unfortunately to prepare chemical super toxic agents is much easier then to make nuclear devices, but the real and especially psychological effect from using them is from some viewpoints bigger. Remember the action of the "Aum Shinrikyo'" group in the Tokyo metro few years ago.

In Russia we have a highly positioned State Commission on Chemical disarmament led by a Presidential nominee, his official representative in Volga region – Mr. Kyrienko. This Commission, of which I am a member, is organizing all the process of the dismantling of CW stockpiles and we are collaborating with US, UK, Germany and other countries.

This August we have opened, and I participated in it, the first line of newly built chemical plant for destruction of Lewisite – one of the oldest CW agents.

This ceremony, held in the presence of several dozens of diplomatic representatives and journalists, showed that Russia is strictly following the approved time schedule for CW dismantling. One of the aspects of all these procedures is to install and keep double, triple and quaternary lines of safety and security control to prevent unauthorized access to any place where CW agents can be found.

To discuss scientific issues of chemical, radiological and biological protection of people (armed forces and population) we created, some time ago at the Academy, an Interdisciplinary Scientific Council which aims to give recommendations to the authorities on the measures that should be undertaken to organize better technical protection control for personnel coming into contact with toxic materials.

Biological Weapons and Bioterrorism

It already took place with anthrax envelopes in US. My friend Professor Lev Sandakhchiev will tell about that more in detail. Here I would like only to point out that our understanding that anthrax, pest, yellow fever, Ebola fever and other bacteria and virus based species are most efficient and widespread dangers in biological warfare seems to be inadequate today.

Much more dangerous could be the products of gene-engineering if instead for instance of vaccines somebody in a modern classical laboratory of molecular biology will purposely manipulate DNA, introducing particular fragments in the genome to make an organism causing lethal illness in a very short time.

That danger is not recognized either by authorities or by public opinion in full scale yet but specialists are already talking about it loudly and our duty is to mobilize intellectual and financial efforts to this absolutely new danger to develop efficient system to identify probable bioterrorists and to develop reliable control for preventing intrusion.

5 VECTOR AND INTERNATIONAL COOPERATION

Lev S. Sandakhchiev
*Director, State Research Center of Virology and Biotechnology VECTOR,
Russian Federation Ministry of Health*

Contribution of Nonproliferation Programs to Strengthening Physical Security and Biological Safety of Work on Dangerous Pathogens

I wish to thank the Organizers of this conference for inviting me to discuss different aspects of ensuring the safe handling of nuclear materials. I have accepted the invitation because the principles, technical solutions, international agreements and the code of practice can be, and must be, largely implemented in handling chemical and, especially, biological agents that represent a threat of the possible deliberate use by terrorists against civilian populations. In my presentation I would only point out to those features of biological agents that cause qualitatively more complex problems in terms of oversight measures towards their storage and inventory.

Let me briefly tell you about our institution and demonstrate results of our collaboration in the nonproliferation area.

I represent the State Research Center of Virology and Biotechnology (VECTOR) of the Russian Ministry of Health, that before 1991 used to conduct defense-related research. Since 1991, we have been implementing a program of restructuring our Center's activities in view of the new economic and political environment. We have integrated into Russia's scientific research programs in public health and veterinary and we are manufacturing products for diagnosis, prevention and treatment of human and animal infectious diseases. We used to receive, and still continue to receive, significant federal support (Russian Ministry of Industry, Science and Technology and Russian Ministry of Health) and assistance from the Regional Administration. It allowed us, apart from implementing scientific research programs, to reconstruct our facilities and become one of Russia's leading manufacturers of public health products.

Regarding the important role of international assistance under the nonproliferation programs, since 1994 VECTOR has been involved in collaborative research programs with ISTC, CRDF, DOE-IPP (USA) and in EU programs. Recently VECTOR has been carrying out so-called ISTC Partner Projects with US DHHS (BTEP/FETP), USDA Agricultural Research Service, and DOD (DTRA, DARPA). The total financial support provided to VECTOR under the nonproliferation programs since 1995 has amounted to US $18.8 million, including US $16.3 million provided via the ISTC. During this period, VECTOR has had 75 international projects (both completed and active), including 49 ISTC projects. It allowed VECTOR to significantly accelerate the implementation of the restructuring program, in terms of obtaining important scientific results, improvement of its telecommunication infrastructure, and meeting current biosafety and physical security requirements supporting the kind of pathogen research we do.

Thanks to integrated support under federal programs of the Russian Ministry of Industry, Science and Technology and the Ministry of Health, and also under regional programs and the international nonproliferation programs, VECTOR managed to adapt itself to the new reality. This can be illustrated by a shift in VECTOR's year 2002 income pattern. The production share in the total income has increased from 22% in 1990 to 66% in 2002. The total income increased from US $2.5 million in 1990 to US $19.6 million in 2002.

While assessing very highly the role of national and international efforts to support VECTOR programs in the nonproliferation area, in strengthening confidence and transparency, I would like to offer several comments on the issues being discuss at this conference:

1. The many years of experience in providing physical security of storing and working with nuclear materials should be fully adapted to ensuring the physical security of both biological agents and

places where research on these is conducted at our institution. As we proceeded with the physical security project, we enjoyed outstanding assistance provided by the Russian Ministry of Atomic Energy (MinAtom) experts in development of designs, selection and supply of devices and instrumentation, construction and adjustment of security systems.

2. We will continue to further enhance the security at our institution, in tight collaboration with MinAtom and our counterparts at DTRA (USA).

3. The accounting of biological materials during storage and transfer can also use the principles and methodology applied to storage of nuclear materials. However, quantitative assessment of biological materials is far more laborious and associated with some subjective factors.

4. A qualitatively more difficult problem is that of keeping an inventory of pathogens during research work since, during such activities, the biological agents, as a rule, grow in quantity and can be represented by not only individual pathogens but also by being present in experiment in the form of infected cell cultures, laboratory animals, etc.

5. Insignificant, hardly accountable quantities of a biological agent, may pose a real threat in terms of uncontrolled leakage of biological material. Unfortunately, this problem does not yet have either an engineering or technical solution. In fact, it is determined by the human factor, i.e. it is necessary to adopt criteria and requirements to personnel to be allowed to work with pathogens, even within the tightly secured laboratory facilities.

6. An associated problem is that highly pathogenic agents are many and they might be accessed during natural outbreaks of disease. Moreover, they can be engineered through simple laboratory manipulations on the non-pathogenic microorganisms available.

I have pointed out these characteristic features of control over the inventory of biological agents and I wish to stress that international collaboration is crucial to solving this problem. To successfully implement the concept of nonproliferation, it is imperative to work to create a principally new atmosphere of trust and cooperation and to demonstrate progress in resolving the outstanding issues of the current nonproliferation programs between Russia, USA, and EU. In my opinion, nonproliferation programs in the nuclear, chemical and biological areas must be definitely integrated on both national and international level. In fact, this is one of the tasks that the Institute for Applied Science (IAS), which I have the honor to represent here as well, has been pursuing during recent years. My colleagues, Academicians Velikhov, Platé and myself – the founding members of IAS – and Mr. John Wood, the Chairman of IAS, will be happy to discuss these issues with you in greater detail.

Thank you.

6 ADDRESS TO THE INTERNATIONAL APPROACHES TO NUCLEAR AND RADIOLOGICAL SECURITY CONFERENCE, 2002

Baroness Thatcher of Kesteven

Introduction

Mr. Chairman, Ladies and Gentlemen. It is an honour to have been asked to attend your reception this evening – though I am somewhat overawed by the expert and distinguished audience I find myself addressing.

But I have at least one advantage. I am part of that rare – indeed almost endangered breed – a politician who began her career in science.

A Serious Business

As the larger-than-life American film producer Samuel Goldwyn once put it in his own inimitable way: 'We've got to take the atom bomb seriously – it's dynamite!'

My friends, if we only could look upon the dangers which now face us with such easy humour. But we can't.

Over the past few days you have brought your expertise to bear in a subject which should command the attention of us all: namely both the security and the potential proliferation of nuclear material.

It is a dialogue between America and Russia, which has been overdue, but we are now forging a partnership, which will bring dividends both for us and for future generations. I pay tribute to you all, but particularly tonight, to our friends from Russia. You serve not only your own nations, but all nations, by your efforts.

Today, we are making up for lost time. But have no doubt, that time must be made up.

The Nuclear Threat

During most of my political life the two superpowers held massive nuclear arsenals, even a small proportion of which could have inflicted untold damage in the event of a nuclear strike. But this knowledge imposed a discipline that made for a kind of stability. The rules were clear, the psychology understood and each side's sticking points known.

But today the threat is different. The proliferation of weapons of mass destruction has fundamentally changed the world in which we and our children live.

We now face the far more menacing risk that such weapons might already be in or soon fall into – the hands of a ruthless and fanatical regime.

Our first line of defence – as this conference has been debating – must be to deter, or to stop such countries from obtaining weapons of mass destruction.

But if our security and watchfulness fails, we must pursue all necessary means either to force these maverick regimes to surrender their weapons, or to neutralize their ability to use them. The choice is that stark.

Wishful Thinking

When I was young there was a popular song that went:

'Wishing will make it so,
Just keep wishing, and cares will go…
And if you wish long enough,
Wish strong enough,
You will come to know
Wishing will make it so.'

But it won't. Though we might wish for the best we must prepare for the worst, and that is where your conference could not come at a more crucial time.

Two hundred years ago, Edmund Burke, one of our country's wisest political thinkers, observed: 'All that is necessary for evil to triumph is that good men do nothing'.

My friends, these words have been echoed down the ages, but they have never seemed truer than today.

7   INTERNATIONAL NUCLEAR MATERIAL PROTECTION AND COOPERATION SITES

Sites of MPC&A Cooperation

MinAtom Weapons Complex

1. Arzamas-16, Sarov, All-Russian Scientific Research Institute of Experimental Physics (VNIIEF)
2. Chelyabinsk-70, Snezhinsk, All-Russian Scientific Research Institute of Technical Physics (VNIITF)
3. Chelyabinsk-65, Ozersk, Mayak Production Association
4. Sverdlovsk-44 (S-44), Urals Electrochemical Integrated Plant (UEIP)
5. Tomsk-7, Siberian Chemical Combine (SCC)
6. Krasnoyarsk-26, Zheleznogorsk, Mining and Chemical Combine (MCC)
7. Krasnoyarsk-45, Zelenogorsk, Electrochemical Plant (ECP)
8. Avangard Electrochemical Plant
9. Sverdlovsk-45, Lesnoy
10. Penza-19, Zarechnyy

Russian Civilian Sites

12. Institute of Physics and Power Engineering (IPPE), Obninsk
13. Bochvar All-Russian Scientific Research Institute of Inorganic Materials (VNIINM)
14. Elektrostal Production Association Machine Building Plant (POMZ)
15. Scientific Production Association Luch, Podolsk
16. Scientific Research Institute of Atomic Reactors (NIIAR), Dimitrovgrad
17. Novosibirsk Chemical Concentrates Plant (NCCP)
18. Khlopin Radium Institute (KRI)
19. Krylov Shipbuilding Research Institute
20. Petersburg Nuclear Physics Institute (PNPI)
21. Joint Institute of Nuclear Research (JINR), Dubna
22. Karpov Institute of Physical Chemistry
23. Moscow Institute of Theoretical and Experimental Physics (ITEP)
24. Moscow State Engineering Physics Institute (MEPhI)
25. Scientific Research and Design Institute of Power Technology (RDIPE)
26. Lytkarino Research Institute of Scientific Instruments (RISI)
27. Beloyarsk Nuclear Power Plant (BNPP)
28. Sverdlovsk Branch of Scientific Research and Design Institute of Power Technology (SF-NIKIET)
29. Tomsk Polytechnic University (TPU)

ARCTIC OCEAN

Laptev Sea

NEW SIBERIAN ISLANDS

East Siberian Sea

Federation

Yakutsk

Lake Baikal

Bilibino

Chersky Range

Kolyma Range

Magadan

Bering Sea

Sea of Okhotsk

Petropavlovsk-Kamchatskiy

U.S.

ALEUTIAN ISLANDS

Khabarovsk

Vladivostok

Sea of Japan

NORTH

PACIFIC

OCEAN

Metropolitan Moscow
— Rivers
— Roads
- - - Railroad
✕ Railroad Station
▲ Airports

Naval Complex

Murmansk Vicinity

30	CBC A1	40	CBC C1
31	CBC A2	41	CBC C2
32	CBC A3	42	CBC C3
33	CBC A4	43	PBZ C1
34	CBC A5	44	PBZ C2
35	CBC A6	45	PBZ C3
36	PBZ A1	46	CBC D1
37	CBC B1	47	CBC E1

Vladivostok region

49	CBC P1	55	CBC P3-2
50	CBC P2	56	CBC P3-4
51	CBC P2-2	57	CBC P4
52	CBC P2-3	58	CBC P5
53	CBC P3	59	CBC P6
54	CBC P3-2	60	CBC P7

Petropavlovsk-Kamchatskiy region

61	CBC K1	67	CBC K4
62	CBC K2	68	CBC K5
63	CBC K3	69	CBC K6
64	CBC K3-2	70	PBZ K1
65	CBC K3-3	71	PBZ K2
66	CBC K3-4		

72 Icebreaker Fleet, Murmansk
73 PM-12 Refueling Ship, Murmansk
74 PM-63 Refueling Ship, Severodvinsk
75 PM-74 Refueling Ship, Vladivostok
76 Navy Site 49, Murmansk
77 Navy Site 32, Vladivostok
78 Navy Site 34, Vladivostok
79 Navy Site 86, Vladivostok
80 Sevmash Shipyard
81 Sergiev Posad
82 Kurchatov Institute

NIS and the Baltics

83 Salaspils Institute of Nuclear Physics, Latvia
84 Ignalina Nuclear Power Plant, Lithuania
85 SOSNY Institute of Nuclear Power Engineering, Minsk, Belarus
86 Kiev Institute of Nuclear Research, Ukraine
87 Kharkiv Institute for Physics and Technology, Ukraine
88 South Ukraine Nuclear Power Plant
89 Sevastopol Institute for Nuclear Energy and Industry, Ukraine
90 Tbilisi Institute of Physics, Georgia
91 BN-350 Breeder Reactor, Aktau, Kazakhstan
92 Kurchatov Institute of Atomic Energy, Kazakhstan
93 Ulba Metallurgical Plant, Ust-Kamenogorsk, Kazakhstan
94 Institute of Atomic Energy, Alatau, Kazakhstan
95 Institute of Nuclear Physics, Tashkent, Uzbekistan

Printed in the United Kingdom by
Lightning Source UK Ltd., Milton Keynes
140249UK00003BB/15/P